THE SCOTTISH
GIN
BIBLE

First published in the UK in 2022
by Black & White Publishing Ltd
Nautical House, 104 Commercial Street, Edinburgh, EH6 6NF.
A division of Bonnier Books UK
4th Floor, Victoria House, Bloomsbury Square, London, WC1B 4DA.
Owned by Bonnier Books
Sveavägen 56, Stockholm, Sweden.

A CIP catalogue record for this book is available from the British Library.

1 3 5 7 9 10 8 6 4 2

ISBN: 978 1 78530 362 3

Layout by Black & White Publishing
Printed and bound in Turkey

blackandwhitepublishing.com

Please drink responsibly.
drinkaware.co.uk

THE SCOTTISH
GIN
BIBLE

100 GREAT GINS FROM SCOTLAND

SEAN MURPHY

BLACK & WHITE PUBLISHING

DEDICATED TO ALL THOSE PEOPLE WHO ARE
THERE WHEN DISTILLERS, GIN PRODUCERS AND
WRITERS NEED A LITTLE ENCOURAGEMENT TO
KEEP WORKING TOWARDS THEIR GOALS.

CONTENTS

*Gin information including prices, botanicals and ABV
are per information supplied by the manufacturer
and are correct at time of going to press.*

FOREWORD

ONE OF THE strengths of gin is that part of its abiding appeal is its ability to endure and adapt through changing and periodically turbulent times. These forced changes come from either economic, political or technological change.

It feels perfectly apt to be reworking a book on the subject as we ease our way through the challenges and adjustments that are posed by one of the largest crises to have been experienced for a generation.

A key asset of the industry in enduring these challenging times is the passion and ingenuity shown by the current crop of producers.

Looking at gin's evolution you can chart our own history from the bad harvests of 1756–57 that did way more to end the gin craze of 1757 than any government intervention, to the invention of the Coffey Still in 1830. This still produced lighter-flavoured spirits, thereby helping to change the motivation for botanical addition from being a job of masking poorly made spirit to being an expression of a distiller's creativity in its own right.

As someone who entered the gin industry in 2008, when the credit crunch was starting to sink its teeth into the economy, I have first-hand experience of not only the meteoric shifts that have gone on within the industry but also the fact that people now drink less but also seek out better quality products.

One of the best parts of my job at the International Centre for Brewing and Distilling is that you have conversations, or read essays, that can truly blow you away. It is a reassuring privilege to realise the calibre of new talent rising up through the ranks.

People of all backgrounds are now being given the chance to make their own mark on the industry, to its continuing strength.

This is also reflected in the emergence of fresh approaches and technology that in generations not so long ago would never have been given the rubber stamp of approval, as new entrants to the market shake things up in a disruptive fashion similar to that of the start-ups within other established markets.

I look forward to seeing what direction we will head in as the Scottish gin industry inevitably evolves as people strive to distil greater meaning and quality into their offerings in what continues to be a crowded market.

It is important to remember that it's a product of sheer human creativity and endeavour you hold in your hand when you sip a long gin and tonic on a balmy summer evening or politely address a gin martini in a cocktail bar.

So it is then, with an open mind and expectant nose, I throw down the gauntlet to the Scottish gin industry to show us what you're made of!

MATTHEW S. V. PAULEY
Gin consultant and assistant professor at the International Centre for Brewing and Distilling at Heriot-Watt University, Edinburgh

AN INTRODUCTION TO GIN

JENEVER JUNIPER: A HISTORY OF GIN

FROM ANCIENT REMEDY to modern drinks superstar, the journey of juniper and its liquid derivatives is a long and illustrious one, with its story flowing from Greece to the Netherlands and Belgium to Britain before reaching America and swinging its way back into Europe and eventually reaching out into the rest of the world.

Gin in its modern form was first enjoyed in Belgium and the Netherlands – the gin we enjoy today is the evolution of a drink that began life as a spirit known as Jenever.

Juniper – or jeneverbes in Dutch – is a common species of evergreen plant – which grows prolifically across the mountains and heathlands of much of the northern hemisphere and Europe. The plants range from small bush-like shrubs up to towering trees, but it's the berry-like female seed cones that have been noted for millennia for their medicinal properties. From these tinctures and remedies, an appreciation of juniper's distinctively fruity, peppery flavour grew.

From the 1600s onwards, the trading power of the Dutch East India Trading Company meant that juniper outgrew its popularity in the Low Countries and went on to be transported around the world, eventually finding its way to England. There it fast became a favourite of soldiers – who, if you believe the stories, dubbed it 'Dutch Courage' after witnessing its nerve-calming and bravado-boosting properties in wars overseas. And, likewise, the nobility took to gin with relish, but with a slightly different agenda – they wished to ingratiate themselves with their newly crowned Dutch-born King William of Orange. Enjoyed by both the toughest and the most indulged members of late seventeenth-century society, cheap, easily produced imitations flooded the streets of London, leading to the creation of the spirit we now call gin.

In the first half of the eight-eenth century – during what became known as the Gin Craze, a period when the juniper spirit spread like wildfire through the city of London – gin was outrageously, and some would say dangerously, popular in England. So much so that William Hogarth immortalised its perils in his 'Gin Lane' print as part of a propaganda campaign that led to the Gin Act of 1751. But while London was awash with gin, it never quite captured the imagination or drinking habits of the majority of Scots in the same way.

However, Jenever, as a distinct drink in its own right, remained a significant import into Edinburgh, through the port at Leith – huge quantities of which were also illegally smuggled into places like the fishing harbour along the shore at Newhaven – thanks to Scotland's trading links with the Low Countries. The tipple soon caught the attention of several notable Scottish figures, such as members of that famous distilling clan the Haigs, and they too dipped their toes into the production of a spirit so very different from our long established national drink. But outside of the capital's walls, uisge beatha – whisky – continued to reign supreme.

Nowadays, considering the vigour with which gin is produced north of the border, you'd be forgiven for thinking that it's a historic Scottish drink. This isn't quite the case, but Scotland has been tied to this wonderful tipple for a very long time, thanks to our nation's rich history of distillation, trade and, of course, innovation. In the past few centuries, many of the spices and herbs used in gin production were brought to Scottish shores via ports such as Leith, Dundee

and Glasgow. At the same time, juniper, which was in plentiful supply in the north-eastern Highlands, was exported to the Netherlands to make Jenever.

However, it's only in the past two decades that gin's popularity has soared in Scotland. In the 1990s, the Scottish company United Distillers & Vintners – the forerunners of Diageo, the world's largest producer of spirits – made the decision to concentrate all their UK spirit production in Scotland. This was a game-changer: gin distillation once again took off in Scotland. In time, this would lead to nearly 70% of the UK's gin being created in Scotland.

The 'small-batch gin wave' – that is, gin's answer to the craft beer movement – began after some quiet experiments by two distillers at whisky company Grant's. The team – master distiller David Stewart, and Hendrick's head distiller Lesley Gracie – wanted to create a premium alternative to that most famous of gin categories, London Dry. The pair were inspired by the successful launch of Bombay Sapphire in 1987, which gin expert Geraldine Coates believes was the true forerunner of a new premium gin sector. The 'lighter style' and striking blue bottle helped to 'revamp the image of the whole category, making it cool and sophisticated'.

Stewart and Gracie hoped to appeal to the modern consumer seeking authenticity and quality in a spirit. The result of

their collaboration was Hendrick's. This gin was something of a slow burner: it launched in 2000 in the US and 2003 in the UK, but it was over a decade before it was recognised as a landmark, pioneering brand. With its kitsch Victoriana marketing, wonderfully accessible taste – and value-for-money price point – and that intriguing cucumber garnish, people slowly woke up to the idea that gin, which had been modestly holding its own in the vodka-dominated spirits market, could be cool again. Such a novel idea wasn't lost on a generation of drinkers who had grown tired of mass-produced alcohol brands and who were drawn to craft and provenance in everything from beer to bicycles. Indeed, the craft-beer market, and its surging popularity both here and in the States, helped to kick-start the growth of the craft spirits market.

Unsurprisingly, it was the whisky industry that first realised the potential of this new market in Scotland. Two of Scotland's expert whisky distillers, Inver House Distillers' Balmenach distiller Simon Buley and Bruichladdich's then master distiller Jim McEwan, began to experiment with white spirits, soon launching the Caorunn and Botanist gins in 2009 and 2011 respectively. Other fledgling producers followed suit and the craft-gin industry was born.

Gin is relatively easy to produce in small quantities. And, as the revival of the juniper spirit took hold, the idea of not only drinking, but also creating gin didn't seem so very outlandish. More and more people began to experiment with the spirit, buying stills and creating their own new gins. Meanwhile, younger whisky companies, established on the basis of the phenomenal popularity of Scottish whisky, started to explore the possibility of using gin as a makeweight while they waited three years for their first whisky to mature in its casks.

This brings us to 2021, and the explosion of gin that's sending waves up and down the country.

Though the pandemic may have hindered or disrupted other industries, the latest figures from the Wine and Spirit Trade Association shows that retail and online sales of gin and flavoured gin rose by 22% in value, breaking the £1 billion barrier for the first time – and that's without the pubs being open.

New distilleries and companies are opening all the time. There are now more Scottish gin distilleries than ever before, which means that now is the perfect time to raise a glass to this heady drink and toast its renaissance in modern Scotland.

GIN STYLE

WHEN YOU ORDER a G&T – or something a little more adventurous – at your favourite local, the range of gins you'll be offered is greater now than ever before.

Even so, all gins must fit within certain parameters. Gin itself is legally defined as a spirit made from neutral spirit of agricultural origin – e.g. the potato and grain beloved of Slavic vodka – and flavoured with juniper berries. European Union regulations state that gin must be bottled with an alcoholic strength of no less than 37.5% ABV – it remains to be seen what will happen to British gin after Brexit – while in the United States it must be bottled at no less than 80° proof (40% ABV).

Creating gin requires artistry, patience and a lot of skill. From gin to genever, these are some of the most popular styles of gin you can enjoy when you order that G&T.

LONDON DRY GINS London Dry Gin is the classic juniper spirit and the best-known style of gin around the world today. The style was developed in the latter half of the eighteenth century, and, as most of its producers were based in London, it became known by the highly original moniker 'London Dry Gin'. However, despite its name, London Dry Gin does not need to be made in London – or even in the UK. The name now refers to any gin products produced under a set of strict EU regulations. These regulations dictate that the base spirit must be a completely neutral spirit, made purely from agricultural sources – meaning it cannot contain any artificial ingredients. After distillation, water and a minute quantity of sweetener can be added, but no additional flavours or colour.

Many distilleries have stayed true to the juniper-led flavour profile of London Dry Gin in their products – Tanqueray and Gordon's to name two of the most famous – but other producers – like St Andrews-based Eden Mill – are more experimental, using the time-honoured London Dry method to produce more unconventional gins with flavours that depart from the traditional.

COMPOUND GINS This style of gin differs from London Dry and other distilled gins in that flavours and botanicals can be added to the neutral spirit without re-distillation. This is sometimes referred to as the 'bathtub method', a term that comes from prohibition-era America, when people would make their own compound gin at home – usually in their bathtubs. In Scotland, Gin Bothy and Orkney are two of the best-known producers of compound gins.

MODERN GINS Modern, or contemporary, gins – which include Hendrick's – have a flavour that is not juniper-led. Juniper is present, but it does not dominate the flavour profile; instead, emphasis is placed more on the other botanicals used – herbal, floral or spicy. This type of gin is becoming more and more popular in the current gin revival.

NAVY STRENGTH GINS The Navy has played an important role in gin's popularity, both in the UK and across the world. Sailors have long had a taste for the drink, and many of the earliest distilleries popped up in maritime cities, with gin being transported all over the world on naval ships.

Navy Strength Gin doesn't have a typical flavour profile that all producers adhere to; rather, gins in this category are united by their high alcohol content. Most have an ABV of 57%, or 100° UK proof. Historically, British naval soldiers tested a gin's strength by pouring it over gunpowder and then trying to light it. If it failed to light or fizzled, they knew the gin was diluted – and therefore unacceptable.

Many Scottish distilleries, such as North Berwick-based NB Gin, produce Navy Strength versions of their classic gin.

GENEVER Dutch Genever, a predecessor to London Dry Gin, is made up of a blend of malt wine and neutral spirit. There are two types of this traditional juniper-flavoured spirit: oude (old) and jonge (young). However, these labels have nothing to do with age, and everything to do with the percentage of malt wine they contain. Oude Genever must have a malt-wine percentage of at least 15% – giving it a malty taste comparable with whisky – while Jonge Genever must have a percentage of no more than 15%. Juniper is the dominant flavour in both, and the inclusion of sugar adds richness to the taste.

OLD TOM GINS Old Tom Gin was popular in eighteenth-century England, a time when distilling practices were far from advanced. Gin was crude, unpalatable – and often dangerous. To mask its poor quality, producers laced it with the sweetest ingredients they could get their hands on, including sugar and liquorice. As a result, Old Tom Gin is often described as the bridge between Genever and Gin – sweeter than London Dry, but drier than Genever. It has more stories alluding to the origins of its name than The Botanist Gin has botanicals – and all fall at different places on the spectrum between the sublime and the ridiculous.

FLAVOURED GIN LIQUEURS These are usually lower in ABV than traditional gins, and tend to be far sweeter and more palatable than regular gins. Brands like Edinburgh Gin have taken advantage of the growing popularity of this gin style, producing a variety of fruity and floral-flavoured liqueurs. Sloe Gin also belongs in this category.

OAK, CASK AND BARREL-AGED GINS These are gins that have been produced and then laid down in casks – mostly ex-whisky in Scotland – to 'mature', for anything from six weeks to ten months. This might sound a little strange. After all, gin is not a spirit associated with barrel ageing. However, it's not a new phenomenon; in the eighteenth and nineteenth centuries, oak was often used to store the spirit, rather than more easily breakable containers. Many producers swear by the process, arguing that barrel ageing imbues the gin with the classic spirit flavours that gin doesn't often contain, especially if the barrel has been used to age another spirit, like whisky, beforehand. As a result, barrel-aged gins often taste woody or smoky, and include notes of vanilla or caramel. Glasgow distillery Makar produces an oak-aged gin – a variant of their original gin.

GARNISHES

FOR ALL THE gins, you'll find garnish suggestions for each one; these are chosen by the gin producers themselves. A good garnish is the perfect accompaniment to a gin-based drink: it will appeal to the eyes, tease with its aroma, and give depth to the drink's taste, accentuating the gin's unique flavour.

These recommended combinations are designed to be a steady, tried-and-tested ground from which to branch out and experiment with different ingredients and flavours. The beauty of this spirit is that it is one of the most versatile, so if you don't like a particular garnish, don't be afraid to switch it out for something you prefer. After all, it's you who will be drinking it (along with your friends, of course) so try out anything you like – it's part of the fun!

The same goes for mixers. Many people assume they don't like gin because they don't like tonic. But while gin and tonic are old pals, they aren't exclusive – other mixers like lemonade, soda and ginger ale make excellent date-mates too.

What is the function of a garnish? Essentially, it's an additive that either complements and accentuates a gin's key flavours, or contrasts and clashes with them. The result is a harmonious balance in flavour, or an unusual combination that will excite and challenge the taste buds.

A GLOSSARY OF GARNISHES

Here are some of the terms you will see used in this book and what they mean.

SLICE Fairly self-explanatory: a finely cut piece of fruit.

WEDGE A thicker segment or slice of fruit.

TWIST AND PEEL A thin slice of the peel of a citrus fruit. Cut using a peeler, with the pith (the fruit's spongy inner lining) carefully removed to avoid added bitterness.

In mixology, the term 'twist' refers to the act of squeezing the oils from the peel by twisting it over the drink before garnishing.

EXPERT'S TIP: to add a visual flourish to your drink, wrap a thin strip of the peel round a paper straw to give it shape, then hang it over the edge of the glass.

SPRIGS OR LEAVES An individual leaf or strand of a herb or plant.

GRATED Using a fine grater, grate a small amount of the chosen spice into the drink.

THE BEAUTY OF BOTANICALS

TRADITIONALLY DESCRIBED AS 'any substance derived from parts of a plant and used in making medicine', the word botanical has taken on a new life of its own within the drinks industry. Now synonymous with gin, it usually refers to the herbs, seeds, roots, berries and fruits that help to flavour the spirit used to create gin.

Each of these botanicals are selected and combined – in much the same way as a recipe – to create a delicate symphony of flavour that will create the character of the final spirit. However, the juniper berries must contribute the predominant flavour if the drink is to be classed as gin. From there on, anything goes when it comes to which additional botanicals are included.

Jamie Shields, of the Summerhall Drinks Lab in Edinburgh, likens the process to making a curry; each person will have their own idea of what constitutes the best recipe – an opinion at least partly driven by the idiosyncrasies of their own taste buds – despite many curries sharing the same three or four defining base ingredients.

And that is the magic of gin, too. Each recipe can be tailored to an individual's preferred tastes and adapted to include ingredients specific to the region in which the gin is made. Perhaps, in time, this will give rise to the gin category gaining its own form of terroir – like Champagne.

In Scotland, it's this sense of provenance and regionality, rather than the mere attachment of a place name to a product, that gives producers their ability to create points of difference in a crowded and competitive market.

And so here is a guide to some of the more prominent botanicals, from the five main traditional options through to some of the more innovative ones used by Scottish gin producers.

FIVE TRADITIONAL BOTANICALS

JUNIPER BERRIES These are to gin what grapes are to wine and malted barley is to single malt whisky. They are the key ingredient that makes gin, gin. Juniper is responsible for that wonderful pine-forest aroma and flavour that makes this spirit so enticing.

Traditionally, juniper was abundant in Scotland. Now, the plants struggle to mature because the country's deer population like to eat them. Most producers now source the berries from places such as Italy, Macedonia, Bulgaria and even Croatia.

CORIANDER SEEDS Sourced in southern Europe, Morocco and India, coriander seeds – not the divisive plant – are often described as the second most important botanical in gin making. These small yellow seeds link particularly well with juniper berries, adding citrus notes and a little spice to the final spirit.

ANGELICA ROOT Native to countries in the northern hemisphere, angelica is found throughout Europe. The roots – and occasionally seeds – are used in gin making to provide a base earthiness. They also help to bond many of the other flavours together.

ORRIS ROOT The root of the iris flower is used in many gins, although the high cost of the powder – a consequence of its labour-intensive processing – means it is used sparingly.

The root is highly prized in the perfume-making industry, thanks to its ability to bond and enhance other – usually lighter – aromas. It's useful to gin distillers for the same reason. Not entirely aroma-free, orris provides a light, sweet floral note said to be akin to the old-school British sweet – the Parma Violet.

CITRUS Citrus provides that clean, refreshing top note in a gin. This sweet, zesty flavour is usually provided by lemon and orange peel – although sometimes

the fruit is used in its entirety. In recent times, a wider variety of citrus fruits has started appearing in recipe lists. Limes, blood oranges, grapefruits and yuzu offer more unusual flavour notes.

OTHER TRADITIONAL BOTANICALS

ANISEED, STAR ANISE AND FENNEL SEEDS Much like liquorice root, these provide different variations of that earthy aniseed flavour.

LIQUORICE ROOT Used mainly as a sweetening agent, liquorice also provides light notes of aniseed.

CASSIA BARK AND CINNAMON These spices are very similar and stem from a common family tree – in fact, cassia is often mis-sold as cinnamon. They both provide a warm, woody spiciness to gin.

GREEN CARDAMOM SEEDS Unmistakably perfumed, these provide a sweet, warming note, with a fresh hint of eucalyptus.

COMMON SCOTTISH BOTANICALS

SEA BUCKTHORN BERRIES Commonly found around Scotland's coasts, this aromatic plant's berries are quite tart and provide a fresh, fruity top note.

ROWAN BERRIES Abundant in Scotland's forests and woodlands, rowan berries have a delicate flavour, which provides a sweet, slightly bitter-tasting note.

BOG MYRTLE A common sight in Scotland, these industrious little shrubs have been used in everything from medicine-making to brewing, and are well known as a great midge repellent. They offer a baseline of resinous notes such as wood and pine.

HEATHER A plant common throughout Scotland, heather provides a grassy, floral note with hints of honey.

GORSE Not heavily used in gin-making, gorse, very common throughout the UK, is said to produce a coconut-like scent when picked.

BLADDERWRACK SEAWEED This distinctive and nutritious seaweed can be found on coastlines around the UK and provides a salty tasting note, which can riff off the more traditional base botanicals.

NETTLES Another common garden weed that has been given a new lease of life as an ingredient in gin making, nettles have been used in brewing and cooking for centuries and provide a fresh, grassy flavour.

ROSE HIPS Better known by Scots of a certain age as 'itchy coos', rose hips are the deep orangey-red fruit of the wild rose. They provide a lovely floral sweetness.

SCOTS PINE NEEDLES Abundant in the ancient forests that dot the country, Scots pine needles have become popular with producers for how they accentuate the pine-y resin-like notes of the juniper.

MILK THISTLE The seeds of this prickly purple flower were traditionally believed to help prevent damage to the liver. The seeds provide nutty oily notes when used to make gin.

RHUBARB Rhubarb stalks dipped in sugar are a traditional treat for Scottish children. When used in gin making, the tasty vegetable adds tart notes that pair well with summer fruits.

THE GINS

SOUTH-WEST SCOTLAND

GLASGOW, ARGYLL & AYRSHIRE, DUMFRIES & GALLOWAY

HILLS & HARBOUR GIN

BOTANICALS
Juniper, noble fir needles, bladderwrack seaweed, mango and peppercorns

PREFERRED SERVE
A 50ml measure of Hills & Harbour Gin over plenty of ice, add some premium tonic and a slice of fresh mango to garnish

PREFERRED GARNISH
A soft garnish like fresh mango

Crafty Distillery, Newton Stewartt
craftydistillery.com
Price £36 / Quantity 70cl / ABV 40%

I N YEARS GONE by, the furthest south you could have found a local gin would have been Girvan, home of Hendrick's, but, as with the rest of Scotland, the south of the country now enjoys its fair share of outstanding gin distilleries and brands.

In Dumfries and Galloway, just south of the Galloway forest park and lying close to the flowing waters of the River Cree, you'll find the home of Hills & Harbour Gin and one of the country's most southerly spirit production sites, Crafty Distillery.

Founded in April 2017 by Graham Taylor, the custom-built distillery has enviable views over the Galloway countryside – and its premium gin is named after the terrain that surrounds the distillery. Filled with forests and unspoiled coastlines, these landscapes offer the 'perfect larder' in which to source ingredients.

The Crafty Distillery prides itself on removing all of what it calls 'distilling jargon' to make their gin and distillery as approachable as possible. This ethos carries through into the way the distillery focuses on accessibility, with the small team offering a range of tours and masterclasses.

Key to this is the unique Galloway Gin Escape, in which guests can explore more of Galloway's stunning scenery while foraging for two of the local ingredients used in the gin. They can then learn how the spirit is made, sample some of the finished product and mix their own delicious cocktails.

Another intriguing aspect to Crafty is that they are one of the few Scottish distilleries to create their own base spirit. Wheat is taken from nearby farms to create a grain spirit that provides a smooth mouthfeel and a delicate sweet flavour to the finished article.

Always innovative, they've not only added limited edition gins and vodkas to their output but also the 'world's first distilled cocktail' which is made using smoked pineapple and burnt oranges, and 100% distilled with fresh fruit.

They recently unveiled plans for their first-ever single malt, Billy&Co, named after Graham's late father, William James Taylor, who helped build the distillery and was something of an expert when it comes to what makes a great dram.

It'll be a community affair, with whisky fans who join Crafty's new members' club given the chance to shape the future whisky by tasting, and giving feedback on, a range of new-make spirit samples.

With its grain spirit created from local

RECOMMENDED COCKTAIL
GALLOWAY FORAGER

Ingredients
50ml Hills & Harbour Gin
25ml grapefruit oleo saccharum
 (made with molasses sugar)
25ml elderflower cordial
Soda water
1 sprig of noble fir, to garnish
1 slice of grapefruit, to garnish

Method
1. Fill a glass tumbler with ice.
2. Add the cocktail mix and top with a dash of soda.
3. To serve, garnish with a fresh sprig of noble fir and a slice of fresh grapefruit.

wheat and foraged botanicals picked to result in a 'smooth and vibrant' gin, Hills & Harbour is ideal to drink neat, with a mixer or in a cocktail.

Chosen to mirror the forests and coastlines that give the area its distinct character, two of the botanicals – noble fir needles and bladderwrack seaweed – are foraged locally. Described as juniper-led with 'hints of forest fir, tropical fruit, spice and a subtle scent of the shore', this versatile gin uses mango to add depth and sweetness, and peppercorns to unify the flavours.

ORO

BOTANICALS
Juniper, coriander, vanilla, orris root, lemon peel, orange peel, cassia bark, Malabar cardamom, pink peppercorns, lemongrass, angelica, fennel, bitter almond, cinnamon, liquorice and one 'forever secret' ingredient

PREFERRED SERVE
Simply, on the rocks or with a Mediterranean tonic. Oro is such a smooth gin that it really doesn't need a mixer

PREFERRED GARNISH
A slice of orange or pink grapefruit

Dalton Distillery, Dalton
orogin.co.uk
Price £35 / Quantity 70cl / ABV 43%

LYING SOUTH OF the picturesque town of Lockerbie and the River Annan, the Oro Distillery was created by the Clynick family in 2017. This idyllic rural setting is so rich with history that it seems almost anomalous to find one of the country's most cutting-edge gin distilleries here. Ray Clynick, head distiller and co-founder, has a background as a scientist and, after graduating from Heriot-Watt International Centre for Brewing and Distilling, he set about working out how he and his family could create a gin based on a 'thoroughly scientific understanding' of how flavour compounds interact with each other.

As part of their research, they designed a bespoke copper still, which they named Bridie – after one of the family dogs – and the result is one of the most distinctive-looking stills you're likely to see. Replete with three partial condensers, the distilling set-up allows Ray and his team to control every part of the process; it's versatile and creates as pure and clean a spirit as possible.

ORO NEGRONI

Ingredients
50ml Oro Gin
45ml Hotel Starlino Vermouth
40ml Campari

Method
1. Fill your mixing glass with ice, add your spirits, stir for 1.5–2 minutes (until fully chilled).
2. Pour through a fine strainer.
3. Serve in a classic tumbler with an ice block and garnish with a twist of grapefruit.

They are also committed to sustainability; the distillery uses locally collecting rainwater – in plentiful supply in Dumfries! – to cool the system.

Unlike much of the distilling industry, the company doesn't produce any waste water. Should you be curious about this fascinating new distillery, then the Clynicks will happily welcome you with a range of tours and tastings designed to show off what the Dalton Distillery is all about, as well as teaching you how to make the 'Perfectly Scientific Gin and Tonic'.

A 'classical style Scottish dry gin', Oro is the distillery's first release. The follow-up, Oro V, is a less traditional gin, which balances the juniper with the lighter, fresher notes of coriander, and uses lavender as a natural smoothing agent.

ORO GIN

Aiming to create a gold standard in Scottish gins, 'oro' means gold in both Spanish and Italian. The Oro logo uses concentric circles to represent the atomic suborbital structure of gold itself.

Oro is created using 15 named botanicals including juniper, coriander, cinnamon, pink peppercorns and vanilla. Ray and the team also use one signature ingredient, which they insist will 'forever remain secret'.

Described as the first-ever 'Scottish Dry Gin' by the Clynicks – owing to the fact that it's made to the strict rules of the London Dry category, but in a Scottish setting – Oro is very clean with its strong base of juniper followed by sweetness from the vanilla and subtle spices from the cinnamon and coriander.

The Hendrick's Gin Distillery Ltd, Girvan
hendricksgin.com
Price £31 / Quantity 70cl / ABV 41.4%

3
HENDRICK'S GIN

TAKE A MEANDER south from Glasgow along the southwest coast of Scotland and you'll eventually reach the stunning vista of the Firth of Clyde and Ailsa Craig. It's here you'll find, close to the coastal town of Girvan, the home of Scotland's pioneering craft gin.

Lighting the torch paper – or firing the still, as it were – of the modern gin renaissance in Scotland, the decision by William Grant & Sons to release what would go on to become the first of Scotland's new craft gins would turn out to be a master stroke.

Growing from minor experiments undertaken by Balvenie Malt Master Dave Stewart and distiller Lesley Gracie to a small release in 1999, Hendrick's has become one of the world's bestselling craft gin brands – topping one million cases sold in 2017. Strange to think that it took so long for the others to eventually catch on.

Gracie, whose ingenuity brought this unique juniper spirit to life, is now head distiller, ensuring Hendrick's is in safe hands. Working with a rare Bennett still from 1860 and a 1943 Carter-Head still bought by Charles Gordon, great-grandson of William Grant & Sons founder William Grant, at auction in 1966, Gracie still creates this hugely popular gin in relatively small batches.

Hendrick's' strength is its market. Oozing Victoriana from every pore, the branding team very cleverly tied the gin to its now legendary garnish, the cucumber, so as to create a distinct serve.

Not to be undone by the rise of even more exaggerated garnishes, Hendrick's have stayed true to their original choice, even going as far as to back the annual Cucumber Day event on 14 June, established in 2011 to highlight the pleasures of this versatile fruit.

The bottle itself is based on an old Victorian era apothecary's vial and features the year 1886 on its label, the year the first William Grant distillery was founded. Events such as an expedition to a South American rainforest to search for a new gin botanical and the creation

BOTANICALS

Orris root, yarrow, juniper, chamomile, lemon peel, orange peel, elderflower, angelica root, coriander seeds, cubeb berries and caraway seeds

PREFERRED SERVE

Hendrick's curious flavour can produce some delightful cocktails The traditionalists prefer a classic gin and tonic, but there are a wealth of alternative preparation methods

PREFERRED GARNISH

Hendrick's should be garnished with that humble yet marvellous green fruit – the cucumber. Three delicate rounds are perfect!

RECOMMENDED COCKTAIL
GARIBALDI SBAGLIATO

Ingredients
25ml Hendrick's Gin
25ml Campari
50ml fresh orange juice
Champagne
Orange wedge to garnish

Method
1. Combine all the ingredients in a frozen or chilled highball glass with no ice.
2. Top with Champagne and garnish with an orange wedge.

of a night of music using cucumbers as instruments have helped to tie the brand to a whimsical vision of Queen Victoria's reign – a major part of Hendrick's' success.

The recent opening of their Gin Palace in 2018 has led to Gracie now having double the number of stills to play with, while other new releases from them have included the Lunar, Midsummer Solstice and the intriguing Orbium.

Lesley Gracie, master distiller at Hendrick's, creates the gin using both of their unique stills to distil two separate heavier and lighter infused spirits which are then blended before the Bulgarian rose petal and specially selected cucumber essence are added post-distillation to create a rich, balanced juniper spirit.

The gin features several classic botanicals such as orris root, coriander seeds and citrus peel, alongside some more unusual ingredients including yarrow, chamomile and elderflower. The result is a gin that's smooth, full of depth and filled with subtle flavours; it's fresh, floral and bottled at a punchy 41.4% ABV.

With a price point that's often cheaper than many of the newer competitors to the market, and an availability that extends from most bar gantries to many supermarket shelves, Hendrick's is easily accessible and considered to be one of the best gins for beginners to find their feet in the category.

BIGGAR

A FAMILY HOLIDAY TO Arran in 2016 served as the inspiration for brothers Stuart and Euan McVicar to start up their own gin business. Over a few tasty gins, the discussion turned what had just been an idea into something they felt was 'definitely doable' due to their shared love of the spirit.

Though both still working full-time as an IT consultant and lawyer respectively, it wasn't long before the idea began to become reality as they started embarking on courses to learn all they could about gin, running a distillery and how to produce this exciting spirit.

A small still soon followed, as well as the embarking upon their first experimentation with recipes, with a list of 32 botanicals selected. Stuart joked that from there it was just a case of keeping going until 'somebody or something stopped them'.

Real help soon beckoned, though, with the team at Strathearn Distillery in Perthshire providing the extra knowledge, expertise and hands-on experience the

pair desired to start the Biggar Gin Co.

They launched Biggar Gin soon after in April 2018, with the first version being produced for them by Strathearn and plans laid to start work on their own distillery if it was the success they hoped.

A gin festival in Glasgow, where they sold around half of their first batch, and a win in the London Dry Gin category in the Gin Guide awards soon followed and they had their answer: they were definitely on to a winner.

Deciding it was time to put their money where their mouth was, Stuart quit his day job in 2019 and started working on Biggar Gin full-time to build the new distillery.

They had until this point been producing Biggar Gin at the Gatehouse Distillery in Auchterarder, where Stuart has one of their two Hoga stills, and the move to their new purpose-built distillery took place in summer 2021 where they now finally have their two stills, a shop and are able to offer customer tours all under one roof.

The Stillhouse, Wyndales Mill, South Lanarkshire
biggargin.com
Price £36 / Quantity 70cl / ABV 43%

BOTANICALS
Juniper, coriander, fresh orange peel, fresh lemon peel, cardamom, cassia, pink peppercorn, lavender

LOCAL BOTANICALS
Rowan berry, nettle leaf, rose hip

PREFERRED SERVE
Lots of ice and either Cushiedoos or Walter Gregors Original tonic

PREFERRED GARNISH
Orange peel with tonic. Lemon peel with ginger ale. Rosemary for something a little more herbaceous

RECOMMENDED COCKTAIL
BIGGAR TEA SMASH
(Anthony Gallant)

Ingredients
37.5ml Biggar Gin
50ml strawberry and raspberry tea
25ml lemon juice
12.5ml sugar syrup
1 bar spoon of raspberry (or
 blackcurrent) jam

Method
1. Place all ingredients in a cocktail shaker with ice and shake for 20 seconds.
2. Pour into an old-fashioned glass with ice and garnish with a raspberry.

Alongside their original London Dry, they've now also created a Navy Strength (the delightfully named Biggar Still) and a limited edition Plum Gin which they create every September and October and celebrates the Clyde Valley's heritage of orchards.

BIGGAR GIN
Using local botanicals such as rowan berry, nettle leaf and rose hip, as well as more traditional ones like fresh orange peel, fresh lemon peel, cassia, and pink peppercorn, Stuart and Euan strived to create a moreish gin that people would always come back to.

Designed to be as enjoyable drunk neat as it was with tonic or in a cocktail, the pair went for a slightly higher ABV to create a richer mouthfeel and a complex finish that's filled with rich flavours including juniper, citrus, light floral notes from the lavender and hints of spice from the cassia.

The result is a fresh-tasting, small-batch gin with a 'Biggar mentality'.

ELLIS NO. 5

BOTANICALS
Juniper, coriander, angelica, oris root, orange peel and lemon peel (the peel of over 140 lemons is used in each batch)

PREFERRED SERVE
Ellis No.5 Lemon Gin is best served with London Essence tonic and garnished with a slice of lemon peel or with a sprig of basil. For those gin drinkers who don't like tonic, try with Bitter Lemon and a slice of lemon

PREFERRED GARNISH
Slice of lemon peel or a sprig of basil

ELJ Drinks, Uddingston
ellisgin.com
Price £31 / Quantity 50cl / ABV 40%

THE ROUTE TO the Scottish gin market, and becoming a successful creator of a gin brand, is not always a straightforward one. While some may launch distilleries or move from other spirits, entrepreneur Carol Jackson began her journey creating ready-to-drink gin cocktails which used what would go on to become her first gin as the base.

Teaming up with Glasgow's Illicit Spirits, she set about launching an exciting range of gins with the first Ellis Gins (named after her daughter), the Scottish Bramble and Butterfly Pea Gins, arriving in 2018. Looking to bring a softer brand to the market, Carol explained that she wanted to add an ethos of fun and colour, and two initial colour-changing gins proved popular before an innovative Pink Shimmer Gin was added in 2019.

Carol was most excited to introduce her Ellis No.5 Lemon Gin. Described as 'sunshine in a glass' the new citrus gin is the ideal summer tipple.

Ellis No.1 – a London Dry – is their most recent release.

6

FORGET ME NOT

ONE OF THE only Scottish gins that comes with true star power, it was created by a team led by *Outlander* actor Caitríona Balfe, who plays Claire Fraser. Caitríona wanted to create a small-batch spirit that would support the arts and 'acknowledge the positive contribution' the creative industries make to society.

Launched at the end of summer 2020, the main botanical for this gin is the eponymous flower (which holds special significance for Caitríona's character's time-travelling adventure in *Outlander*) as well as rose hip, elderflower and, interestingly, beetroot.

With the arts sector suffering during the pandemic, a portion of the profits will go to support arts charities.

With a recipe developed by a group of drinks experts and production taking place at Perthshire's Strathearn Distillery, Forget Me Not Gin is designed to be a floral gin that features a juniper base combined with delicate notes of citrus and hints of coconut, lavender and elderflower.

BOTANICALS
Juniper, coriander seed, liquorice root, angelica, forget-me-not flower, rose hip, elderflower, lavender, orange peel, beetroot, coconut

PREFERRED SERVE
Over ice with a premium tonic and a slice of orange

PREFERRED GARNISH
A slice of orange

Lucky Lion Ltd, East Kilbride
forgetmenot.com
Price £45 / Quantity 70cl / ABV 40%
Contract Distilled by Strathearn Distillery,
Perth & Kinross, Scotland

SOUTH-WEST SCOTLAND

Inspirited Distillery, Strathaven
inspirited.co.uk
Price £34 / Quantity 70cl / ABV 57.1%

INSPIRITED

FEATURING WHAT IS perhaps the most unique foray into the world of producing gins, the team at Inspirited set out to offer a completely different gin experience.

Founder Lawrence Nicholson had the idea for Inspirited back in 2019 while enjoying a drink and discussing the popularity of gin with his family at New Year.

Deciding that he wanted to set out to offer customers the chance to design their own gins, he then faced the problem of how to come up with a process that created the same quality, consistency and replicability of any major distillery but on a single bottle-by-bottle basis.

Working for six months with a specialist Canadian company ABT, Nicholson designed a unique new still in October 2019.

Not only did this special still allow Inspirited to achieve their goal of being able to provide a vast range of quality custom gins, it also gave them the scope to explore a huge range of botanicals and create their own family of gins.

And experiment, they did, getting so technical that they even got to the stage of cracking open cardamom pods and putting in individual seeds just to see how they affected the final recipe.

This led to the creation of three core gins: Inspirited Original, a light and accessible citrus-forward gin; Summer Berry and Hibiscus, an extremely popular pink gin made using only natural flavours and ingredients; and the deliciously complex Navy Strength Spiced.

Giving gin fans the chance to be their own recipe makers, Inspirited offers a selection of 35 botanicals across six different flavour categories on their website, as well as some truly excellent tools to help aspiring gin makers to create their very own gin.

BOTANICALS

Juniper, coriander, cardamom, orris root, galangal root, kaffir lime leaves, galangal, baobab, grains of paradise, Kentish cobnuts, angelica root, pink peppercorns, cassia bark

PREFERRED SERVE

Plenty of ice, Fever-Tree Indian tonic, a slice of grapefruit or orange

PREFERRED GARNISH

A slice of grapefruit or orange, or a little chilli

RECOMMENDED COCKTAIL

SOUTHSIDE

Ingredients
7 mint leaves
60ml of Inspirited Navy Strength
  Spiced
25ml of lime juice (freshly squeezed)
15ml sugar cane syrup

Method
1. Shake all ingredients with ice.
2. Strain into a chilled glass.

INSPIRITED NAVY STRENGTH SPICED

With their Navy Strength, which comes in at a powerful 57.1%, Lawrence explained that the plan was to produce a gin that works at a higher strength, and not just a gin with a higher ABV.

Deciding on a spicier profile, they settled on a recipe that included some intriguing botanicals such as Kentish cobnut – which has the taste and consistency of a hazelnut but the texture of coconut – and galangal root which provides a lovely citrus flavour that helps to balance out the spice of the pink peppercorn and grains of paradise.

One of those rare gins that is smooth and rich enough in flavour to enjoy neat over ice, it also doesn't lose any of that flavour when mixed with a bolder tonic.

GLASWEGIN

BOTANICALS
Juniper, coriander, angelica, milk thistle, orange and chamomile flower, pink peppercorns and bay leaves

PREFERRED SERVE
Keep it simple with Glaswegin, a no-nonsense tonic, and a few cubes of ice

PREFERRED GARNISH
A slice of green apple or a sprig of mint

Illicit Spirits Distillery, Glasgow
glaswegin.com
Price £39 / Quantity 70cl / ABV 41.1%

EVERYBODY LOVES WORDPLAY and one of the most fantastically named gins around must be Glasgow's own Glaswegin. This stylish gin was founded by Andy McGeoch after he noticed that most of the gins in Glasgow's bars were not from the city itself.

Aiming to create a no-frills gin representative of its home city, Andy worked with Darran Edmond of Illicit Spirits. This was a nod to Glasgow's industrial past, as Darran's distillery is under a railway bridge close to Glasgow's Central Station. It was important to Andy and his team that every element of Glaswegin reflected the city and the people that inspired it.

Coming in a stunning square white bottle, with unique wraparound branding, designed by Glasgow School of Art graduate, Paul Gray, Glaswegin's recipe focuses on the 'magic eight' botanicals, the most notable of which include milk thistle, orange and chamomile flower, pink peppercorns and bay leaves, to create a gin that's as enjoyable to visit as the city in which it is made.

SOUTH-WEST SCOTLAND

Burnbank Farm, Strathaven
mcleansgin.co.uk
Price £39.50 / Quantity 70cl / ABV 39.4%

McLEAN'S GIN

STARTING OFF IN a small cupboard in a tenement in Glasgow's southside, the entrepreneurs behind McLean's Gin quickly earned the title of Scotland's smallest gin producers and offer what could surely be one of the most unique gins in this book.

Unofficially beginning when founder Colin's mother- and father-in-law gave him a gin-making kit for Christmas in 2015, the home brewing enthusiast eventually got round to using it as a surprise gift for his now-wife Jess and after trying it out, the idea that would go on to become their now hugely popular gin line was sprung.

Experimenting with compound gin production – a process that doesn't involve re-distillation and sees the neutral spirit steeped with the botanicals over a set amount of time – the aspiring duo soon hit upon using Colin's small-business experience in the construction industry to launch a gin brand.

Using their own name – it was 'suitably Scottish sounding', according to Colin – the new firm began with a tiny

budget and first produced their now discontinued Signature Gin in 2017, and after an article in a local newspaper, they hit national news leading to an explosion in interest (and orders).

Before a move to bigger premises in Strathaven in June 2018 (the same week they got married), Jess came up with the novel idea of Something Blue, a wedding inspired Scottish gin.

Making 150 bottles of the exciting new blue-tinged gin, it became one of their fastest sellers, selling out in just five days.

> **BOTANICALS**
> Tonka bean, buchu leaf, juniper, orris root and clitoria ternatea
>
> **PREFERRED SERVE**
> Serve over ice with 1 part Something Blue and 2 parts Bon Accord tonic
>
> **PREFERRED GARNISH**
> Pink grapefruit or mint and cucumber

SOUTH-WEST SCOTLAND

Now their core gin, the pair produce all of their compound gins in small batches over 48 hours with their main line including Floral, Citrus, Spiced, and Cherry Bakewell gins, with plans for an exciting new gin in the near future.

FRENCH 75

Ingredients
1 tbsp lemon juice
1 tsp sugar syrup
50ml gin
Champagne
Ice
Lemon zest
1 tsp Blue Curacao

Method
1. Pour the lemon juice, sugar syrup and gin into a cocktail shaker then fill up with ice.
2. Shake well then strain into a Champagne flute.
3. Top with a little Champagne, leave to settle (as it will bubble up) then fill up with more Champagne.
4. Swirl gently with a cocktail stirrer then garnish with a strip of lemon zest.

SOMETHING BLUE

Discovering the complex delights of the South African botanical buchu leaf, Colin was looking for a recipe to use this exciting ingredient with its notes of blackcurrant and mint – however this was before wife Jess fell in love with the flavours of the tonka bean on a date night at one of Glasgow's most prestigious restaurants where they were using it to flavour their ice cream.

Focusing on a 'marriage' of their two new favourite botanicals, Colin combined them with six other botanicals including juniper, orris root and clitoria ternatea (proposed 'ter-nay-shiah'), better known as butterfly pea flower which gives the gin its unique colour.

Colin describes the gin as being 'like no other', as the cold compounding process doesn't extract the dry juniper notes you'd expect, resulting instead in floral notes of lavender and violet, as well as the rich complexity of the tonka bean, buchu and the slight spice of cardamom.

Bottled at the slightly lower strength of 39.4%, Colin explained that it's 'strong enough to get you through your wedding day but not strong enough to make you forget your speech.'

PENTLAND HILLS GIN

CELEBRATING HOGMANAY IN 2015, Phil and Tabatha Cox decided it was time to stop everyone else running their lives for them and take up something they were actually passionate about.

Which is how they then found themselves 18 months later developing an exciting new gin named for the scenic hills close to their base in South Lanarkshire.

Setting up within the grounds of their home in a converted stable block, the pair – along with their chocolate Labrador and mascot Panza – have built a gin distillery that is based on quality, not quantity, creating the gin in a manner that has a minimal impact on the environment.

The Pentland Hills Original Gin was launched in November 2018. Awards soon followed and, with a focus on sustainability, the Pentland Hills team also unveiled a refill scheme, which sees their customers returning their empty bottles to the distillery to be cleaned, labelled, refilled and reused.

The team grow their own botanicals where they can (including Scottish

BOTANICALS
Juniper berries, pink peppercorns, angelica, coriander, cardamom, orange, mint and cocoa nibs

PREFERRED SERVE
Simple tonic (no big flavours) and a slice of pink grapefruit or orange

PREFERRED GARNISH
A sprig of mint

Tarbraxus Distillery, South Lanarkshire
pentlandhillsgin.com
Price £42.99 (Refill Price, £32.99)
Quantity 50cl / ABV 58%

SOUTH-WEST SCOTLAND

DOGGYJITO

(serves 2)

Ingredients
100ml Pentland Hills Gin
10 mint leaves
4 wedges of pink grapefruit
50ml sugar syrup
100ml apple juice

Method
1. Place the mint in a mortar and pestle and gently press until you can smell the mint fragrance being released from the leaves.
2. Add 2 pieces of the grapefruit and gently squish.
3. Transfer to your glass and then add the sugar syrup onto the concoction and stir.
4. Fill your glass until it is half full with crushed ice and swirl the ice and mixture together, then add your PHG and apple juice.
5. With a long swizzle stick, give the glass another good stir and garnish with the sprig of mint and the grapefruit (give this a small squeeze before placing into the glass).

juniper) and power their systems through renewable energy – even using the tails from their distillation process to make hand sanitiser. This commitment to sustainability has seen them recognised by the influential Gin Guide as one of their most environmental and sustainable distilleries for 2021.

The newest addition to their range, the Navy Strength, was launched in the middle of 2020, and is in honour of Phil's grandfather, who served aboard the destroyer *HMS Electra*. A portion of the profits are donated to the Royal Naval Benevolent Trust. Already proving popular, it's already gained recognition at some prestigious awards. Not bad for a distillery that produces all of 500 litres a year!

PENTLAND HILLS

Phil double distils each batch of this dynamic gin on his 60-litre copper still, which is named Amy after the famous female pilot Amy Johnson (all of the names for their stills have an aviation theme), for extra smoothness, and it is a delightfully approachable spirit.

Each bottle features a beautiful etching of both Panza and the stunning hills that give the gin its name, while the recipe which includes orange, mint and cocoa nibs offers an appealing mix of flavours that will have you looking forward to your next refill.

CLYDESDALE GIN

HAVING BOUGHT THEIR family farm in 2017 at Climpy in South Lanarkshire, Jenny McKerr and her husband had assumed that beef and sheep farming, which was their passion, would be what would be keeping them busy for the foreseeable future.

Little did they know then, they'd also soon be indulging in not only their passion for drinking gin but also creating one of their own just a year later.

After a tasting session in the capital with some friends led to the decision to buy a still and make a real go of it, Jenny soon found herself experimenting with botanicals and creating recipes for what would go on to become their first release in May 2018, Drovers Gin.

Taking a different approach to many gin makers, in looking at how their gin would pair with food, specifically a steak dinner, and joking that the farm was really good for growing weeds, the 'Gin Farmer' decided to combine these farm-grown Scottish botanicals – heather and thistles – with traditional citrus botanicals

and more savoury ingredients like allspice and pink peppercorn to create a gin that would match well with a nice cut of beef.

The resulting recipe features 13 botanicals and took several attempts to get perfect and find the right balance before Jenny was happy that every sip of the new Drovers gin brought a different botanical to the fore.

Keen to create a range that would cater for a wide range of palates, Jenny then followed up Drovers with the Farmer's strength, which is bottled at a whopping 57% ABV and features artwork from local artist Anne Anderson, and the Drover Artist Edition, which features a beautiful portrait of the Drovers Gin coo by talented Ayrshire artist Jan Laird.

Their latest offering, the Clydesdale Gin, which celebrates their regional heritage of Clyde valley fruit growing and of course the Clydesdale horse, arrived in November 2019.

The family also offer the ideal getaway for any gin fan with their own

GOOD HONEST SCOTTISH GIN

ARTIST EDITION

Clydesdale
GIN

Bottle Number 700 | Distiller *Jenny McKerr*

A refreshing blend of local rhubarb and apples with traditional gin botanicals.

40%vol
50cl℮

Distilled in Climpy · Forth · Lanarkshire Scotland

The Wee Farm Distillery, South Lanarkshire
theweefarmdistillery.co.uk
Price £33 / Quantity 50cl / ABV 40%

Juniper, citrus, cardamom, cassia, rhubarb and apples

PREFERRED SERVE
Delightful over ice with elderflower tonic, a slice of apple and a mint leaf

PREFERRED GARNISH
Apple or mint, or go wild and have both

RECOMMENDED COCKTAIL
CLYDESDALE BREEZE

Ingredients
50ml Clydesdale Gin
50ml apple juice
100ml elderflower tonic

Method
1. Add all of the ingredients to a mixer with ice.
2. Shake for a minute before straining into a glass with ice.
3. Garnish with apple and mint.

on-site gin-inspired luxury cottage that not only comes with a cosy wood burning stove but also its own hot tub.

CLYDESDALE GIN

Produced on their 50-litre copper still, lovingly named Morag II (the successor to the original Morag, a 30-litre) all of Jenny McKerr's gins are still tiny batches that are handcrafted on-site at her farm.

Whereas their original gin, Drovers, was designed with savoury foods in mind, the Clydesdale gin is a much sweeter affair. Infused with garden-grown rhubarb and sweet apples, it is ideal for a more adventurous drinker with a sweet tooth.

The gin takes light crisp notes from the fresh apple and a delicate finish from the rhubarb to create a refreshing spirit that's delightfully drinkable.

CROSSBILL

BOTANICALS
Scottish juniper and rose hip

PREFERRED SERVE
Serve with ice and 1 part gin to 2 parts
Fever-Tree tonic

PREFERRED GARNISH
A twist of orange

Crossbill Highland Distilling, Glasgow
crossbillgin.com
Price £38 / Quantity 70cl / ABV 43.8%

NAMED AFTER THE UK's only endemic bird species, the Scottish crossbill, this thrilling brand brings a little bit of the Highlands to the east end of Glasgow.

Beginning life in Aviemore in 2012, the operation moved to the east end of Glasgow in 2017. Increasing production to match demand for their popular gins, they established the Crossbill Gin School alongside the distillery, which gives masterclasses in gin production and allows visitors to distil their own gin on-site.

Originally set up by an experienced spirits producer, Crossbill use wild Scottish juniper to make gin – something that was once historically common across the UK – reverting back to a time when the country's juniper stocks were a thriving industry, so much so that the berries were exported to the Netherlands to make genever in the seventeenth century.

Today, the juniper harvest in the Highlands only occurs in small batches once the juniper plant is mature (three years old) to ensure the sustainability of the wild crops. To honour this, Crossbill created the small-batch Crossbill 200, which features juniper sourced only from a single 200-year-old tree.

The pioneering gin brand has also added a limited edition pineapple gin liqueur and a sumac gin to their range. This exciting expression is created using staghorn sumac, a botanical found in New Hampshire, the natural habitat of the Scottish crossbill's cousin, the North American red crossbill.

CROSSBILL GIN

Described as a celebration of juniper and rose hip, two 'bold and fresh' Highland botanicals, Crossbill is made in small batches and aims to reflect the terroir of the Scottish juniper. Unusually, it only uses these two botanicals, distilled in a ratio of 60:40 to ensure the juniper takes the lead when it comes to the final spirit, which results in a real gin lover's gin.

RECOMMENDED COCKTAIL

GARDEN MARTINI

Ingredients
45ml Crossbill Gin
25ml Pincer Botanical Vodka
15ml lime juice
A dash of elderflower cordial
½ sprig of rosemary
5 basil leaves
2 lime wedges
A bay leaf to garnish

Method
1. Muddle the lime, basil and rosemary.
2. Add the Pincer, Crossbill, elderflower cordial and lime juice.
3. Shake and double-strain into a chilled martini glass.
4. Garnish with a bay leaf.

Illicit Spirits Distillery, Glasgow
illicitspirits.co.uk
Price £35 / Quantity 70cl / ABV 40%

ILLICIT

FOUNDED IN 2017 by International Centre for Brewing and Distilling graduate Darran Edmond, Illicit Distilling was the result of his decision to cut out on his own after working as a distiller for other gin companies.

With a desire to pursue his favourite facet of the profession – new product development – Darran decided to set up his own distilling business so he would be free to play around with new recipes and ideas.

A distinctly modern and forward-thinking set-up with a 'uniquely lo-fi, urban approach', Darran's approach is simple and based around a philosophy of old-meets-new; he explained that he loves to take old processes, equipment and recipes, and tries to spin them into something unique and contemporary.

As well as making several exciting gins and a single-cask rum, Darran is never afraid to experiment and even found a novel way to recycle the botanical by-products from his gin distilling process to use them to create an inno-

BOTANICALS
Juniper, coriander, angelica, liquorice root, orris root, pink peppercorns, rose hips, tonka beans, orange peel (fresh), nasturtium (vap. infused)

PREFERRED SERVE
A ginger beer highball – simply add to tall glass with ice then top up with ginger beer

PREFERRED GARNISH
A sprig of mint

vative beard balm. He mixes them with argan oil, shea butter and beeswax to create a natural balm that 'enhances and nourishes beards', which is massively popular with bearded drinkers.

Perhaps most excitingly, the innovative distiller's experiments recently led him to 'create a smoked gin' using juniper berries smoked over Scottish peat.

Dubbed 'Blacklist', the unique gin was

SOUTH-WEST SCOTLAND

43

black in colour and featured a recipe of botanicals designed to enhance the smoky flavour, including black cardamom, black pepper, allspice, bitter orange and Lapsang Souchong tea.

Showing off his skill and desire to pursue new ventures, he also expanded into rum production in 2019, producing a hand-distilled spirit from 100% organic molasses. Each cask of this exciting spirit is aged exclusively on-site in the distillery's railway arch distillery.

ILLICIT NEW TOM

Sitting alongside the Illicit London Dry and the Blacklist Smoked Gin, the idea for the New Tom Gin sprang from Darran's desire to create a modern take on 'Old Tom'. Trying to replicate the style using atypical ingredients, he put a recipe together with tonka beans, pink peppers, nasturtium petals, and rose hips, before finally sweetening the spirit post-distillation with bee pollen and honey.

Focusing on balance, Darran explained that they wanted something which was obviously sweeter than a London Dry but still very much a gin, with subtle honey, fruit and floral notes.

Produced by hand in small batches using a direct-fired copper pot still, the gin is a gateway for people who have come to the spirit through gin liqueurs and are looking to move on to something closer to the real thing, reflecting the new and evolving palates of modern gin drinkers.

River Clyde, Glasgow

NEW TROPICAL

Ingredients
35ml New Tom Gin
25ml pink grapefruit tepache*
10ml sugar syrup
2 drops saline
Coconut oil

As a big fan of the indigenous Mexican drink tepache, Darran says it makes the perfect addition to this cocktail.

Method
1. Mix all the ingredients together.
2. Add ice, strain and serve straight up.
3. Garnish (possibly ironically) with a little umbrella, and a spritz of coconut oil for added texture.

*To make the tepache
1. Chop a pineapple into chunks (rind and all, to ensure you include all the natural yeasts growing on its surface).
2. Mash up the chunks and place in a clean glass jar with some water and sugar for a couple of days before straining off the resulting liquid. In this variation, he also includes a couple of thick slivers of pink grapefruit.

LENZIE GIN

BOTANICALS
Juniper, cranberry, bilberry, lime, orange, oris root, cassia root, coriander seed, liquorice root, angelica root

PREFERRED SERVE
Tonic and ice

PREFERRED GARNISH
Freeze-dried cranberries

Lenzie, Glasgow
lenziegin.co.uk
Price £37.50 / Quantity 70cl / ABV 42%

NOT CONTENT WITH running a multi-award-winning shop and deli, which they opened in November 2012, husband-and-wife team Mark and Sue Billington set out to create a delicious gin that they could stock on their deli's shelves.

Born from their love of the juniper spirit, the pair wanted to design a modern take on a traditional gin that would reflect their local area, an idyllic little town on the outskirts of Glasgow, and fit alongside the other artisan and locally sourced products they sold.

With advice coming from the 'The Friends of Lenzie Moss', a volunteer group for the local nature reserve, and expertise supplied by expert distiller at Distillutions, Lewis Scothern, the pair settled on a recipe that included local botanicals cranberry and bilberry.

The hard work paid off when their exciting new gin launched in June 2018. A surprise Scottish Gin of the Year award win just a year later at the Scottish Gin Awards picked Lenzie Gin out as the best of the Gold Medal winners in a blind tasting session with an expert panel of judges (which included yours truly).

Mark and Sue have just launched their first gin liqueur, also made using their key botanicals cranberries and bilberries. It is great over ice, with prosecco and in cocktails.

The label is designed by owner Sue and features the iconic Queen's Building in Lenzie, home to the Billington's shop, and illustrations of the local berries used to make this exciting gin.

Fully deserving of its award win, Lenzie Gin is a great example of how to do a gin right, with flowery notes on the nose combined with that incisive juniper and the sweet tart hints of the cranberry and bilberry to deliver a fruity finish that makes for a delicious and refreshing G&T.

LENZIE MOSS

Ingredients
60ml (about 2 shots) Lenzie Gin
2 to 3 dashes of Angostura Bitters (according to personal preference; we prefer 3)
Lemonade
Ice
Lemon peel, to garnish

Method
1. Place 2 to 3 ice cubes in a glass along with the Angostura Bitters.
2. Swirl the glass to coat the inside of the glass with the Bitters. The ice will dilute and help with this process. Do this for roughly 20 to 30 seconds, or until the glass is coated a slight orange colour. Pour out the ice and any diluted water into the sink.
4. Add 3 to 4 ice cubes, depending on glass size and size of ice. Add the Lenzie Gin and top with lemonade.
6. For garnish, take a piece of peeled lemon and squeeze it over a flame to express the oils onto the glass. Gently rub the lemon peel around the rim of the glass and drop onto the cocktail.
7. The garnish stage is not necessary; however, it does add a whole new dimension to the drink thanks to the smell of the lemon oils.

SOUTH-WEST SCOTLAND

MAKAR

BOTANICALS
Coriander, angelica, lemon, liquorice, black peppercorns, cassia bark and rosemary

PREFERRED SERVE
In a glass with ice and a good-quality tonic water

PREFERRED GARNISH
Mild green chilli

The Glasgow Distillery Company, Glasgow
glasgowdistillery.com
Price £28 / Quantity 70cl / ABV 43%

THE SCOTTISH GIN BIBLE

SCOTLAND'S BIGGEST CITY is home to the Glasgow Distillery, a production site that can be found in the Southside at Hillington Park. It not only produces its awarding-winning gin Makar, but it is also Glasgow's first independent single malt whisky distillery, founded in 1902.

Named after the ancient Scots word for 'the maker or craftsman', the flagship gin's moniker reflects the creativity and craft that's invested so carefully in its production.

This unassuming distillery hidden away in an industrial estate in Hillington is a spirits marvel. It might be small, but it produces no fewer than five styles of gin, two botanical vodkas, one blended malt, one spiced rum and a core range of three single malt whiskies. Established in 2012 by drinks industry veterans Liam Hughes, Mike Hayward and Ian McDougall, the Glasgow Distillery Company saw their unique distillery built just two years later in 2014.

Three of the site's five German-made

stills are named after close members of the three founders' families: Tara and Mhairi are named after Liam's daughter and Ian's daughter respectively. The third, 'Annie' – named for Mike's great-grandmother, who introduced him to gin and instilled in him a passion for making the juniper-led spirit – is used to create their distinctive gin.

The remaining two stills are named after Glasgow's most influential female artists, sisters Margaret and Frances Macdonald who were prominent members of 'the Glasgow Girls', a group of pioneering female artists who had an undeniable influence on forging the iconic 'Glasgow Style' of art which the city still celebrates to this day.

Forward-thinking and innovative, the Glasgow Distillery Company have a young distilling team, half of which are female, with a keen focus on the Glasgow community and key partnerships with local businesses such as Glasgow airport. This synchronised outlook means that Makar gin is the first that many will meet on their arrival into the city.

MAKAR GIN

When it was first released in 2014, Makar marked another first for the company: it became the first gin to be distilled in Glasgow.

Its distinctive seven-sided bottle immediately draws the eye and, of

---

RECOMMENDED COCKTAIL
DRY MAKAR MARTINI

Ingredients
50ml Makar Gin
15ml dry vermouth
A dash of orange bitters
A twist of orange or lemon peel to
 garnish

Method
1. Pour the Makar, vermouth and orange bitters into a glass and stir for 20 to 30 seconds.
2. Strain into a classic cocktail glass.
3. Add the twist of orange or lemon peel to garnish.

---

course, there's symbolism in its design with each side representing one of the seven carefully selected botanicals chosen to combine with juniper to make the distinctive Makar Gin.

Angelica, liquorice, black peppercorns and cassia bark are combined with juniper berries in the pot of Annie – the copper still – before lemon, coriander and rosemary are vapour infused in the helmet to create a gin whose heavy rosemary overtures give it a big vibrant flavour that really stands out from the crowd.

SHOOGLE

BOTANICALS
Kumquat, sweet orange, tonka bean,
cassia bark, orris root, angelica root,
coriander seed, pink peppercorn and
dried baobab fruit

PREFERRED SERVE
A premium tonic water

PREFERRED GARNISH
Halved kumquats or a slice of orange

Shoogle Spirits, Glasgow
shooglespirits.com
Price £38 / Quantity 70cl / ABV 42%

THE WORD 'SHOOGLE' is defined by the family-run firm behind this playful gin as 'a distinctly Scottish wobble', and it's this sense of fun that they wanted to convey when designing their signature spirit.

Chris Payne and his family, after travelling around Scotland visiting other gin makers at festivals, markets and distillery events, decided to put their passion for Scotland's 'other spirit' to the test by making their own.

Faced with changing circumstances because of the 2020 lockdown, and fed up with watching Netflix and doing Zoom pub quizzes, the decision to create Shoogle became a priority, and they began to work together to refine a recipe that they all loved.

Settling upon the traditional London Dry Gin style, with a preference for a juniper, citrus and spicy gin, they felt that it would be both time consuming and potentially cost-prohibitive to build their own facility straight away, particularly given the marketplace during the pandemic.

Instead, they turned to Lawrence Nicholson at Inspirited, a local distiller who was more than happy to work with them to help them scale up their recipe and realise their goal. The family now have their own distillery in Glasgow and are currently working on a number of seasonal gins and other spirit expressions to follow on from the success of their first product.

SHOOGLE GIN

Working with a local greengrocer in Glasgow, they tried a variety of citrus fruits before settling on kumquats, a delicious fruit that provides a citrus sweetness that doesn't overwhelm the more subtle flavours of the gin.

Combined with a healthy dose of juniper, as well as other vibrant botanicals like tonka bean, pink peppercorn and dried baobab fruit, it makes for a textured gin with the citrus sweet and piney notes leading then followed by a warming spice finish.

THE SHOOGLE

Ingredients
3 to 4 fresh kumquats
60ml Shoogle Gin
15ml Elderflower Liqueur
15ml Rosemary Syrup
10ml lemon juice

Method:
1. Muddle the kumquats in the bottom half of a cocktail tin.
2. Shake together with ice in a cocktail shaker.
3. Strain into a Collins glass with a nice chunk of ice.
4. Top up with soda water and garnish with four kumquat halves and a sprig of rosemary.

SOUTH-WEST SCOTLAND

Loch Lomond Distillery, Lomond Estate
benlomondgin.com
Price £33 / Quantity 70cl / ABV 43%

BEN LOMOND

INSPIRED BY THE majesty and beauty of the scenery surrounding one of Scotland's most famous mountains, Ben Lomond, the distillery behind this adventurous gin looked to the hills and slopes of Loch Lomond and the Trossachs when putting their recipe together.

The plan? To create a gin that invokes the adventurous spirit of Ben Lomond, which as the most southerly Munro, provides ready access to a sense of discovery and adventure for the many Scots who visit it.

When looking for ideas for the ingredients for their fledgling juniper spirit in 2019, brand manager Alisha Goodwin – along with an experienced gin consultant – profiled the flora that surrounded the distillery and the national park. Finding an abundance of rowan berry trees, she began to harvest the berries herself to experiment with what would become Ben Lomond Gin's first key botanical.

Looking to create a classic London style gin (with a Scottish twist), they sought out another Scottish ingredient with a sweetness that balanced out some of the heavier tangier notes from the rowan – blackcurrants. The blackcurrants deliver a punchy sweetness that contrasts with the rowan berries and still allows the juniper room to breathe.

It wasn't long before the team had settled upon a final recipe of 11 naturally occurring botanicals including Szechuan pepper, fresh orange and rose petals.

The resulting spirit 'Gateway to the Highlands' became the first-ever premium gin released by the whisky specialists at the Loch Lomond Distillery.

The Ben Lomond team also offer two flavoured gins available all year round: the Ben Lomond Blackberry & Gooseberry Gin and the Ben Lomond Raspberry & Elderflower Gin.

Contract distilled by the Glasgow Distillery, the Ben Lomond Gin team works hand-in-hand with their distillation partners, leading the liquid development and continuous involvement

Juniper, coriander, angelica root, cassia bark, orris root, liquorice, rowan berries, blackcurrants, Szechuan pepper, fresh orange and rose petals

PREFERRED SERVE
4:1 Fever-Tree Indian Tonic Water

PREFERRED GARNISH
Fresh blackberries or raspberries, complimented with lime

RECOMMENDED COCKTAIL
HIGH ROAD

A deliciously sweet, citrusy cocktail that showcases the natural flavouring of Ben Lomond Gin.

Ingredients
40ml Ben Lomond Gin
20ml lemon juice
15ml cassis
Fever-Tree Indian tonic
A wedge of fresh lemon

Method
1. Fill a highball with cubed ice.
2. Add Ben Lomond Gin, lemon juice and cassis.
3. Top with Fever-Tree Indian Tonic, garnish with a wedge of fresh lemon.

with the day-to-day distillation process. The final liquid is produced, bottled and labelled by hand in Loch Lomond Group's bottling facility located on the Ayrshire coast of Scotland.

BEN LOMOND GIN

The first thing you'll notice about this vibrant gin is the eye-catching bottle, and, unsurprisingly, there's a reason for that. Designed to evoke the spirit of the Munro that gives this spirit its name, the height and majesty of the bottle are used to convey the strength of the iconic landmark, while the wooden closure has been added to connect drinkers to the woodland of the Trossachs were some of the spirit's unique ingredients hail.

The rich blue colour of the bottle is designed to be reminiscent of the winter's day on the loch, while in sunlight, it changes to a vibrant lilac-purple as a nod to the summer heather on the hills.

Designed to be accessible to a variety of palates, thanks to the distillery's existing international audience, there is something in there for everyone – from the sweet notes of the blackcurrant, citrus notes from fresh oranges, to the spice of the Szechuan pepper and the floral tones of the rose.

A hugely versatile spirit, that's perfect for experimenting with, it's the ideal gin for any would-be explorers keen to try something new.

KINTYRE

I T MIGHT NOT be the easiest distillery to get to, but the road trip becomes wholly worthwhile when you see the scenery. Set within the idyllic region of Kintyre and the Torrisdale Castle Estate, Beinn an Tuirc boasts magnificent views over the Kilbrannan Sound to the Isle of Arran.

The estate, comprising around 1,200 acres of hills, forests and farmland, has been in the Macalister Hall family since 1872; the distillery itself is housed within a former piggery on the estate. The unique rural setting is not the only asset that enables Beinn an Tuirc to stand out. These gin producers are also renowned for their focus on green energy and sustainability.

The hill that gives the distillery its name – Beinn an Tuirc is Gaelic for 'hill of the wild boar' – provides the water source for the hydro-electric scheme that powers their German-imported copper still, rather delightfully named Big Don. Launched in 2016 by the estate's laird Niall Macalister Hall, wife Emma

BOTANICALS
Juniper, coriander, angelica root, orris root, cubeb berries, almond, bitter orange peel, cassia bark, liquorice, lemon peel, Icelandic moss and sheep's sorrel

PREFERRED SERVE
In a glass with Fever-Tree Mediterranean tonic

PREFERRED GARNISH
A sprig of basil

BASIC GARNISH
Fresh mint

and family dog Crumble, this wonderful distillery has now launched a café and gin school, with three eco-cabins to accommodate weekend guests.

Firmly rooted in the local community, Beinn an Tuirc have pledged to plant a tree in their dedicated wood-

Beinn an Tuirc Distillery, Campbeltown
kintyregin.com
Price £36 / Quantity 70cl / ABV 43%

land area for every case of Kintyre Gin sold in a bid to minimise their impact on the environment. They also invest a percentage of their profits in community projects and local business start-ups. In addition to their signature Kintyre Gin, the distillery has created a range of other products including Kintyre Pink Gin, using Scottish raspberries, and the award-winning Tarbert Legbiter Navy Strength Gin, a spicy and flavoursome spirit named after a Viking sword.

KINTYRE GIN

Created by head distiller Su Black using spring water from a well found on the estate, Kintyre Gin has 'delicate floral notes and a citrus body, with a juniper and spice finish'. It uses, among other botanicals, juniper, coriander and liquorice as well as Icelandic moss and sheep's sorrel, both of which have been chosen to reflect the countryside that surrounds the distillery.

GINTYRE

Neil says, 'The Gintyre marries a sophisticated combination of sweet grenadine syrup with the aromatic citrus notes of Kintyre Gin.'

Ingredients
35ml Kintyre Gin
30ml grenadine
2 tsps pomegranate seeds,
    plus extra to garnish
Soda water

Method
1. Add the Kintyre Gin to a cocktail shaker followed by the grenadine.
2. Throw in the fresh pomegranate seeds and shake over ice.
3. Pour into a sugar-frosted martini glass, top up with soda water and garnish with more pomegranate seeds.

THE GINS

SOUTH-EAST SCOTLAND
EDINBURGH & THE LOTHIANS, THE BORDERS

19
LILLIARD

BOTANICALS
Rose hip, rowan, meadowsweet, juniper, angelica, liquorice, orange peel and elderflower

PREFERRED SERVE
Fever-Tree Naturally Light Tonic

PREFERRED GARNISH
Slice of orange peel

Lilliard Ginnery, Jedburgh
bornintheborders.com
Price £35 / Quantity 70cl / ABV 40%

THE ROLLING AGRICULTURAL lands of the Scottish Borders provide a bountiful supply of amazing foodstuffs and beautifully made craft products, but, up until a few years ago, you would have struggled to find any kind of local spirit or beer to slake your thirst.

Now the region has come alive with some exhilarating new businesses launching gins, laying down whisky spirit in casks and investigating the possibilities of even more expansive drinks like rum.

One such distiller, which narrowly missed out in the race to become the first to produce gin in the region in centuries, is Lilliard. This small craft producer is firmly embedded in the Borders region, specifically the Teviot Valley.

Launched in February 2017, the gin is named for the heroine of the Battle of Ancrum Moor, fought just outside the village in 1545. The tale tells how Lilliard took up her sword against Henry VIII's army and fought to the death to avenge the murder of her lover at the English army's hands. Part of the Teviot Valley, where the gin is based, now bears her name, and it seemed only fitting that this gin should be named in honour of her indomitable spirit.

Based at Born in the Borders, side-by-side with the successful brewery there, the Lilliard Ginnery uses locally foraged botanicals which are chosen to reflect the vibrant rural landscape. Given its eponymous heroine, it's interesting that the team has disavowed the traditional naming conventions – in the interests of gender equality – to name their still after a man, with Donald being their chosen moniker (and, no, not that Donald).

They've recently introduced Ruby, the latest addition to their growing family of spirits, which is an enticing ruby gin liqueur made with Scottish raspberries and strawberries and their base gin. Ideal for adding to prosecco, it's also been designed to enjoy neat over ice, with soda or in a refreshing summer cocktail.

As a fully interactive location, the Born in the Borders ethos is also reflected in the Ginnery where you'll find micro-tours and gin classes.

RECOMMENDED COCKTAIL
MAIDEN'S PRAYER

Ingredients
1 measure Lilliard Gin
1 measure triple sec
1 tsp orange juice
1 tsp lemon juice
A twist of lemon to garnish

Method
1. Shake all of the ingredients over ice.
2. Strain into a cocktail glass.
3. Garnish with a lemon twist.

SOUTH-EAST SCOTLAND

The Borders Distillery, Hawick
thebordersdistillery.com
Price £33 / Quantity 70cl / ABV 43%

KERR'S

PLANS FOR AN exciting new distillery project in the Borders were first announced back in 2017, and just a year later the first whisky distillery in the region in nearly two centuries (since 1837, in fact) saw spirit flow from its stills.

The Scottish Borders will soon have its very own single malt whisky thanks to the Borders Distillery founders and drinks industry veterans John Fordyce, Tim Carton, Tony Roberts, and George Tait.

First up, though, on their spirit journey, was the creation of a novel gin redistilled from their own malted barley base spirit, with every grain used grown within 30 miles of the distillery. Kerr's is a key signifier of how exciting their upcoming whisky range will be, sharing the same DNA via that exciting new-make spirit.

The gin is named after William Kerr, a native of Hawick, and one of the nineteenth century's most accomplished botanists, who travelled widely across Asia sending back hundreds of plant specimens to the famous Kew Gardens.

Not long after the arrival of their gin, the Border's Distillery team announced the launch of another new spirit, Puffing Billy Steam Vodka. Instead of using regular filtration – which would strip away the 'delicious flavours' and 'soft textures' of the barley – the spirit is instead steamed through charcoal inside Puffing Billy, the rare Carterhead still which gives the vodka its name.

The distillery is an award-winning conversion of the former Edwardian electrical works building, and houses the four copper pot stills, the Carterhead, and the retail shop.

Celebrating the traditional manufacturing that Hawick is famous for, and featuring signature architectural elements like its stunning glass roof, it's another glowing example of why spirits tourism in Scotland is flourishing. The distillery's popular tours allow drinks fans to peek behind the curtain at how the distinctive spirits range is made.

It'll still be a few years before you can try the distillery's single malt, but in

SOUTH-EAST SCOTLAND

BOTANICALS

Juniper, coriander, liquorice root, angelica root, orris root, cassia bark, almond, and sweet and bitter citrus peels. The malted barley base is also a key component of Kerr's distinctive character

PREFERRED SERVE

Fever-Tree Mediterranean tonic water and lots of ice

PREFERRED GARNISH

A slice of orange

RECOMMENDED COCKTAIL

KERR'S FLORA

Ingredients
50ml Kerr's Gin
25ml lime juice
25ml raspberry syrup
2 dots of Jerry Thomas or Angostura
 Bitters
Ginger ale

Method
1. Shake gin, lime juice, bitters, and raspberry syrup with ice.
2. Strain into an ice-filled Highball glass.
3. Top with ginger ale.
4. Garnish with a lime wheel and fresh raspberries.

the meantime they offer two intriguing blended whiskies in Clan Fraser Reserve, a traditional blended Scotch, and Lower East Side, a blended malt.

KERR'S GIN

Focusing on key long-term partnerships with local farmers gives the distillery team complete control over the creation of their spirit, from barley to bottle.

The only Scottish Gin made with malted barley spirit in a Carterhead still, Kerr's is created by gently steaming the botanicals in the spirit vapours, rather than boiling them as with most other gin stills.

The resulting spirit both celebrates the rich body and mouthfeel of the malted barley spirit and the subtle flavours infused from the botanicals.

21
1881 HONOURS GIN

THE BORDERS IS a region that has grown in confidence when it comes to distilling over the past few years; largely neglected for over a century or so, it's now one of the most exciting places to find new and up-and-coming gin distilleries.

One such example is that of the 1881 Distillery, recently opened in one of Scotland's most famous hotels, Peebles Hydro. Named for the year in which the iconic hotel was opened, the distillery aims to take advantage of the site's famous water supply which helped make its name as one of the most famous spa resorts in the UK during the hydropathic craze which swept the country in the Victorian era.

The Shieldgreen spring in the nearby Tweed Hills, known then for its legendary purity, which was said to be able to alleviate a wide range of ailments, is now the same water that flows from the stills of the new distillery and is used by the hotel to make an exciting new gin range.

Built on the site of what was the hotel's

BOTANICALS
Juniper, cassia bark, birch bark, coriander, grains of paradise, milk thistle, angelica root, cardamom seeds, fir needles, grapefruit peel, bay leaves, hawthorn berries

PREFERRED SERVE
Over ice with a splash of tonic

PREFERRED GARNISH
Pink grapefruit peel or a wedge of orange or ginger

1881 Distillery, Peebles
peebleshydro.co.uk/1881-gin
Price £39 / Quantity 70cl / ABV 57%

SOUTH-EAST SCOTLAND

The River Tweed at Peebles

original swimming pool, the distillery also has its own gin school. Thought to be the UK's largest residential, it features glass jars full of botanicals, including those sourced within the grounds of the hotel itself, and 26 individual copper stills.

Hugely popular with the hotel's guests and non-residents it allows them to design and distil their own unique bottle of gin under the guidance of 1881's gin team, with the Hydro team stating that these gin school experiences have become popular

for gin enthusiasts, corporate events, wedding parties and other events.

According to gin distiller Dean McDonald, each of the four gins in their line tells a part of the distillery's history so that they can share their knowledge and heritage with visitors.

As well as the 1881 Honours Gin, the distillery also offers Hydro Gin London Dry, Pavilion Pink, and the intriguing Rafters Gin Subtly Smoked, crafted by blending oak-smoked water into the gin during the proofing process.

HONOURS GIN

Honours Gin – the 1881 Distillery's Navy Strength gin, at 57% ABV – is produced in the 150-litre copper Arnold-Holstein still, which is named Felicity after the daughter of Peebles Hydro general manager Patrick Diack.

It is made using locally foraged botanicals such as birch bark, milk thistle, fir needles and hawthorn berries, which all grow abundantly in the Tweed Valley.

HONOURS NEGRONI

Ingredients
25ml Honours Gin
25ml Campari
25ml sweet vermouth

Method
1. Add the gin, vermouth and Campari to a mixer with ice and stir until it feels cold.
2. Strain into a tumbler over fresh ice.
3. Garnish with a slice of orange.

NB Distillery Limited, North Berwick
nbdistillery.com
Price £32.50 / Quantity 70cl / ABV 42%

NB GIN

ALONG THE COAST from the capital, on the shoreline of Scotland's outstanding golf coast, lies the seaside town of North Berwick, and the home of one of the country's most successful gins.

Founded in 2013 by husband and wife partnership, Steve and Vivienne Muir, the NB Gin team moved into a new distillery and luxury visitor attraction in 2018, reflecting the success the brand has had since its formation. Their gin is now exported across the Atlantic to the United States.

This reach is a far cry from their origins as a 'kitchen ginnery' with Viv and Steve experimenting in their home with a 'pressure cooker and some old central heating pipes' to create a range of recipes in their pursuit of what they describe as the perfect London style gin.

Distilled using only eight botanicals – Viv claims they only need those eight to create a classic gin – NB won a silver medal at the Gin Masters awards at the first time of asking and, even more impressively, was voted the 'Best

BOTANICALS
Juniper, coriander seed, cassia bark, orris root, cardamom, lemon peel and grains of paradise

PREFERRED SERVE
Serve with NB Tonic over ice and garnish with a peel of orange or a slice of pink grapefruit

PREFERRED GARNISH
Peel of orange or a slice of pink grapefruit

London Dry Gin' at the World Drinks Awards in 2015.

NB are regular sponsors of the BRIT Awards after-party, but it's not just music royalty who've noticed the quality of their gin. Indeed, the Queen chose NB as the only gin brand to appear in her official commemorative publication for her ninetieth birthday celebrations.

RECOMMENDED COCKTAIL

B'S KNEES

This cocktail was created especially for the BRIT Awards.

Ingredients
50ml NB Gin
25ml lemon juice
25ml B's syrup

For the B's syrup
250ml bees' honey
125ml hot water
2 rosemary sprigs
5g crushed black peppercorns

Method
1. First create the syrup by mixing the ingredients together and allowing them to cool.
2. Then add all of the gin, lemon juice and syrup to a shaker and shake over ice before double-straining into a coupe glass. No garnish required.

NB Gin now comes in two editions: the classic London Dry, and NB Navy Strength London Dry, a navy-strength gin at 57% ABV. Recently, what is described as the 'world's first London Dry citrus vodka' was also added to the NB core range.

The Muirs and their team designed their new distillery to be energy neutral, with solar panels providing much of the power and a special system built to capture rainwater, store it and allow it to be reused for the condensing process of stilling. And now, situated at Halfland Barns, near the fourteenth-century Tantallon Castle, the 2018 opening of the NB visitor centre will bring fans closer to the gin than ever before with bespoke tours filled with sampling, learning, luxury and even canapés by award-winning local chef, John Paul McLachlan.

NB GIN

As a London Dry Gin, NB is a true traditional gin lover's gin with big hits of juniper followed by clean notes of citrus and a slight spice that adds a nice complexity.

Bottled at 42% ABV, it works as well neat over ice as it does with a mixer or as the base spirit in a cocktail. The Muirs have also teamed up with Bon Accord soft drinks to create their own tonic water designed as the perfect complement to NB gin and their spirits range.

RULE

FOLLOWING IN THE footsteps of their forebear Andrew Usher – one of Scotland's preeminent whisky pioneers and the 'father of whisky blending' – and his descendants, husband-and-wife team Julie and Colin McLean set up the aptly named Bloodline Spirits to continue the rich distilling heritage of Julie's family.

Though the family have a rich connection to Edinburgh, where they created their famous whiskies, the pair moved the story to the borders with the next chapter in a new distillery in Peebles.

Taking the name from a river that runs through the Usher's former estate of Hallrule House near Jedburgh, Rule Gin is produced on their copper still, lovingly named Christa after Julie's great grandmother.

A classic London Dry it features exotic botanicals in citrus fruit pomelo and the raspberry-like Nordic cloudberry. These intriguing ingredients combine with traditional botanicals like liquorice and sweet orange peel to create a gin that is delicate but saturated with delicious flavours.

BOTANICALS
Juniper, coriander seed, angelica root, grains of paradise, liquorice, pomelo, cloudberries, sweet orange peel

PREFERRED SERVE
Fill a Copa glass with plenty of ice and a premium tonic

PREFERRED GARNISH
A slice or twist of blood orange. For a sweeter flavour add an elderflower tonic with pink grapefruit

Bloodline Spirits Ltd, Peebles
rulegin.co.uk
Price £36 / Quantity 70cl / ABV 42%

SOUTH-EAST SCOTLAND

Fidra Fine Spirits Ltd, Ballencrieff
fidragin.com
Price £39 / Quantity 70cl / ABV 42%

FIDRA

IN 2017, CLOSE neighbours and friends Jo Brydie and Emma Bouglet were looking for a change of direction career-wise.

Drawing inspiration from not only the stunning East Lothian coastline and landscape that surrounded them, but also that of the story of the all-female team at Lussa Gin, they began the steps that led to them creating a delicious, locally foraged dry gin.

Based at Ballencrieff near the picturesque town of North Berwick, they decided to buy a 5-litre still, dubbed Sadie after its former owner's mother-in-law, to develop their recipe with the aim of creating a coastal gin that reflected the area in which it was made.

Having got really into their recipe development, Jo and Emma soon realised they were getting distracted by trying to use too many botanicals. They decided instead to focus on a smaller selection with an emphasis on the botanicals they could forage around them or grow in the stunning scenery of East Lothian.

BOTANICALS
Juniper, rose hip, sea buckthorn, elderflower, lemon thyme, thyme

PREFERRED SERVE
Plenty of ice and Walter Gregor's tonic

PREFERRED GARNISH
A slice of lemon thyme or a sprig of thyme

A chance encounter with Walter Micklethwait of Inshriach Gin at the Scottish Gin Awards in 2017, helped them to fine-tune and scale up their production at his shed distillery near Aviemore before they finally launched in September 2018. A lasting partnership, Walter continues to make their gin at Inshriach using their botanicals.

The gin is named after a tiny, but beautiful island off the coast close to North Berwick in the Firth of Forth.

SOUTH-EAST SCOTLAND

KIDNAPPED
(Lewis & Clarke Artisan in Gifford)

Ingredients
50ml Fidra Gin (infused)
12.5ml elderflower cordial
70ml fresh apple juice
Squeeze of lime juice
Eteaket Loose Leaf Blooming
 Marvellous tea

Method
1. Make a cold brew by adding 2-3 dessert spoons of Eteaket Loose Leaf Blooming Marvellous tea to a bottle of Fidra Gin and leave for 35-40 minutes until the gin takes on a green tea look, then strain.
2. Then mix the infused gin with a squeeze of lime juice, the elderflower cordial and fresh apple juice.
3. Shake over ice in a shaker and strain into a Martini glass, finishing with a sprinkle of Blooming Marvellous.

Fidra, which has a working Stevenson lighthouse, is said to have been the inspiration for Robert Louis Stevenson's *Treasure Island*.

Swinging from pirates to smugglers,

the area around Yellowcraig is also historically known as a safe spot for Dutch boats coming under cover of night to drop off their illicit cargo to be taken ashore.

A truly small-batch gin, just 250 bottles are produced from each run meaning the pair of entrepreneurs have been able to grow the business steadily while still maintaining their day jobs. Following several awards and appearance on 'Secret Scotland' with Susan Calman, they've been growing their reputation as one of Scotland's most exciting up-and-coming craft gins.

FIDRA GIN

Coming in a stylish, tall clear bottle, with a label that displays a map of the beautiful island for which it is named, Fidra Gin contains a variety of hand-picked and homegrown botanicals, including rose hip, sea buckthorn and lemon thyme.

The foraged sea buckthorn brings a light saline note, while the lemon thyme, which they grow themselves, offers a more subtle citric note than you'd find in more traditional botanicals like lemon or lime peel.

Finally, the locally sourced elderflower rounds out the gin with sweet floral notes, creating a coastal gin that's as rewarding as any jewels or coins you'd find on a treasure island.

25
THE COLLECTOR

HAVING ALREADY ESTABLISHED a top brewery in 2004, Steve and Jo Stewart were looking at new ideas for the brand when Jo decided to pursue her own passion project. A huge gin fan, she was keen to create a spirit that not only sat alongside their award-winning beer range but also paid homage to the history of the local area. The resulting gin was inspired by botanist and local legend George Forrest, with Jo working with the botanical specialists at the Secret Garden Distillery to pay homage to Scotland's 'Indiana Jones of the plant world'.

One of the first western scientists to explore Yunnan in China he is thought to have collected over 30,000 specimens on his travels. Jo included two of those discoveries in their recipe.

Featuring a striking 'peel and reveal label' The Collector Gin is made using seven botanicals, including a dwarf day lily (one of Forrest's most famous discoveries). It's a nicely rounded gin that has a balance of floral, peppery and fruity notes that flow well with the juniper.

BOTANICALS
Juniper, hemerocallis forrestii, sorbus forrestii, angelica root, winter savory, lemon verbena

PREFERRED SERVE
Mediterranean tonic and ice

PREFERRED GARNISH
Blueberries

Bilston Glen Ind Estate, Loanhead
stewartbrewing.co.uk
Price £37.50 / Quantity 70cl / ABV 41%

SOUTH-EAST SCOTLAND

The Pentland Still Ltd, Midlothian
kingshillgin.com
Price £38.99 / Quantity 70cl / ABV 44%

KING'S HILL

LYING TO THE southwest of Edinburgh, the Pentland Hills are almost as much of a walker's paradise for the city's residents as Arthur's Seat, offering superb views and a way to rediscover nature away from the hustle and bustle of the crowded streets.

It was on one of these jaunts in the hills that inspiration struck for local entrepreneur Alexander (Sandy) Morrison, who was taken by the wealth of botanicals growing there right under his nose.

With classic gin ingredients such as sloes, gorse, rose hips and rowan berries, as well as a few patches of juniper, growing in abundance, the fledgling producer decided to make something using all of the wonderful ingredients on his doorstep.

Pouring over some maps of the Pentland Hills to mark out good foraging spots, he was hit with his second light bulb moment, discovering the ideal name for his new spirit when he came across one particular site.

The King's Hill piqued his interest

BOTANICALS
Juniper, coriander, angelica root, orris root powder, cassia bark, liquorice root, rose hips, orange peel, lemongrass, heather tips, elderflower, gorse flower

PREFERRED SERVE
Mediterranean tonic and plenty of ice

PREFERRED GARNISH
Large slice of grapefruit or orange

and after some strenuous research, he discovered a story to match that of the small-batched gin he hoped to produce.

The regally-named location was named for a famous incident. King Robert the Bruce was celebrating his return from exile with fellow nobleman, Sir William 'The Crusader' Sinclair of Rosslyn.

Sir William claimed he had seen a white stag on the King's estate and bet that he could capture the rare prize

KING'S HILL BRAMBLE FIZZ

Ingredients
50ml King's Hill Gin.
50ml soda water (or Mediterranean tonic if you prefer)
25ml freshly squeezed lemon juice
12.5ml sugar syrup (equal parts water and sugar mixed together)
12.5ml Creme de Mure (Chambord is a good alternative if you can't find it)

Method
1. Fill a cocktail shaker with a handful of ice and add the gin, lemon juice and sugar syrup.
2. Shake together well and strain into a highball glass with fresh ice.
3. Stir in the soda water.
4. Slowly add the Creme de Mure so it 'bleeds' through the drink.
5. Garnish with a bramble (or blackberry) and a slice of lemon.

before the King did. What was at stake if he lost the bet? His head.

In return, the Bruce wagered his entire Pentland Estate as his own stake. Safe to say, the roguish Sinclair was able to save his own neck and claim the prize, with the king even getting to see the hunt finale and celebrate with his rival. 'The Crusader' named the spot where he was victorious in Robert's honour, the King's Hill.

Now based in Loanhead, around six miles southeast of Edinburgh, Sandy creates his gins on an Iberican pot (lovingly named Marion after his late grandmother) still at the foot of the Pentland Hills. He released Kings Hill Gin in 2018.

KING'S HILL GIN

In summer 2020, Sandy redesigned his classic gin bottle to feature a striking blue hue (representing the clear water they use from the Glencorse reservoir) and copper embellishments to create a more contemporary vessel worthy of carrying his popular and royally named gin.

Championing the area in which it is made, Kings Hill is classic style London Dry Gin that lets the hand-picked Scottish botanicals shine. The gorse, heather and elderflower offer stunning floral notes, reminiscent of a spring walk in the Pentland Hills.

LINGIN

FOLLOWING OVER THREE decades in the IT industry, Alyson and Ross Jamieson decided to pursue their life-long love of the spirits industry (encouraged by a gin or two) by launching a new distillery in their hometown.

The birthplace of Mary Queen of Scots, Linlithgow had a strong distilling heritage that ended in the early 1980s when the last of its five distilleries closed its doors. Aiming to recapture that rich history, the enterprising pair decided to bring spirit making back to the royal burgh in 2016.

Following over a year of research and development, they decided to enlist the help of some of the local community to perfect what would be their first release. The response they received was overwhelming: over 300 people replied to their social media post, volunteering to help. Whittling this number down to 119, the team spent 2017 working on a recipe for a London Dry Gin and released their first product LinGin, in January 2018.

The duo then decided to create a new line of gins named for Mary Queen of Scots' four ladies in waiting – the Four Marys.

Made in consultation with a local historian, Bruce Jamieson, and their taster community, the new range featured the Bonny Bramble, Zesty Sherbet, Subtly Spiced and Forever Fresh Gins, with each bottle featuring the story of these fascinating characters.

Faced with the ongoing pandemic lockdown in 2020, Alyson and Ross decided to create a colourful new gin to show their support for the NHS and bring some summer holiday colour and flavour to people locked at home. The new Colours range aimed to 'bring the colours, scents and flavours of the beach to your door'.

The gins were designed to be enjoyed on their own or mixed and matched to create 15 different flavour combinations – the Yellow (Yuzu), Green (Lime), Pink (Raspberry) and Blue (Coconut) gins all come in at 40% ABV. Distilled at Linlithgow Distillery using the same base

Linlithgow Distillery, Linlithgow Bridge
linlithgowdistillery.co.uk
Price £38 / Quantity 70cl / ABV 43%

recipe, the new gins feature 8 of the 12 original botanicals before being infused with their predominant flavour and bottled by hand.

Following the success of their current range, The duo have just released a new cask-aged gin and a Navy Strength gin and have plans for a whisky

LINGIN LONDON DRY

Originally developed on their smaller still Gleann Iucha (Gaelic for Linlithgow) and produced on their 500-litre stainless steel G-still, dubbed Scotty (after the 'Star Trek' character Montgomery Scott who is due to be born in the town in 2222), their LinGin London Dry is created using the locally sourced Meadowsweet, which gives it its unique flavour and grows on the nearby banks of the canal and Linlithgow Loch, as well as more traditional botanicals like cubeb, cardamom and bitter orange.

The beautiful new bottle features a striking design that is both tactile and reminiscent of the town's skyline and palace and the LinGin itself is a juniper-prominent gin with lively bursts of citrus and vanilla, and a warm finish.

Designed to be a very smooth and versatile gin, it is ideal for mixing with various different types of mixer for a range of tastes and mouthfeel.

BOTANICALS
Meadowsweet, cubeb, bitter orange, cardamom and cassia along with one secret herb, one secret spice and one secret pepper

PREFERRED SERVE
Over ice with Fever-Tree Mediterranean tonic

PREFERRED GARNISH
Orange slice or sprig of fresh rosemary

RECOMMENDED COCKTAIL
SOUTHSIDE

Ingredients
50ml LinGin
25ml lime juice
12.5ml sugar syrup
Dash of Angostura bitters
8 mint leaves

Method
1. Shake and strain into glass over crushed ice.
2. Garnish with a sprig of mint.

28

ACHROOUS GIN

BOTANICALS
Juniper, coriander, orris, liquorice, angelica, fennel seed and Szechuan peppercorns

PREFERRED SERVE
A double over lots of good ice, with Fever-Tree Naturally Light Tonic to taste

PREFERRED GARNISH
Though not a fan of garnishing a G&T, Porteous says a slice of lemon or grapefruit works well with the spirit if you're keen to add some fruit

Electric Spirit Co., Edinburgh
electricspirit.co
Price £34.99 / Quantity 70cl / ABV 41%

I N LEITH YOU'LL find one of Edinburgh's most exciting up-and-coming gin producers.

A one-man show, the Electric Spirit Co., now based at Tower Street, was founded by MSc-educated distiller James Porteous at the end of 2014, with the brand's original micro-distillery in Leith following in 2015.

The rich history of the port of Leith can be traced back to 1544. From the 1770s onwards it flourished and by the Victorian era was one of the most important docks, trade hubs and industrial centres in Britain. It's long had a connection to both the trade of gin and its creation, and re-establishing those historic links lies at the heart of what Porteous and the ESC aim to do.

The idea for the gin's branding began with some scribbles based on a simplified image of a light bulb. Porteous says this represents the innovative, entrepreneurial ideas that are the brand's distinctive hallmark, adding that he chose the electric name to reflect the thoughts around the vivid designs and flavours that are the founding principles of his company.

The spirit of partnership that has become so prevalent in the drinks industry is another element that the ESC founder is keen to develop, leading him to create gin collaborations with top drinks events around the country.

The ESC's relationship with Edinburgh's Timberyard restaurant has seen the creation of a series of single-botanical distillates using gorse flower, lemon verbena and Scots pine, all distilled on a rotary evaporator.

Having recently expanded its capacity, the Leith-based company has created a lasting partnership with the newly founded Port of Leith Distillery and was among only a small selection of gins from across the globe to score 96 points or higher out of the possible 100 available at the International Wine and Spirits Competition (IWSC) in 2019.

The first thing you'll notice about Achroous Gin is the striking neon orange bottle, a branding that seems a touch ironic given that the name of the gin springs from the ancient Greek word for colourless or achromatic. However, Porteous is quick to explain that this refers to the gin and not the bottle.

To make the Achroous, the Leith-based distiller uses juniper and coriander as well as other traditional ingredients such as orris, liquorice and angelica root, alongside two more unusual botanicals in the form of fennel seed and Szechuan peppercorn. Where the former provides a 'lovely anise sweetness', the latter adds a 'woody, floral aroma' and an 'almost sherbet-like citrus on the tongue'.

Each of these botanicals has been chosen to create a tangible impact on the character of the finished spirit.

RECOMMENDED COCKTAIL

ACHROOUS NEGRONI

The producer says, 'I love the way it makes a negroni taste. You need bold flavours to avoid getting lost in a glass with Campari and the vermouth, but I think a good whack of juniper and the character of the fennel seed and peppercorns really shines through.'

Ingredients
25ml Achroous Gin
25ml vermouth rosso
25ml Campari
1 orange twist

Method
1. Pour the gin, vermouth rosso and Campari into a glass, mix well.
2. Strain into a glass with ice.
3. Garnish with a twist of orange peel.

Crabbie's Distillery, Edinburgh
Price £23.44 / Quantity 70cl / ABV 40.2%

CRABBIE'S 1891

FEW SCOTTISH GIN makers can claim a drinks heritage as illustrious as the Edinburgh-based producers Crabbie's Gin range.

The range was inspired by the original recipes of John Crabbie, a pioneer of the Scottish spirits industry. Crabbie established himself as a greengrocer and a drinks merchant in Leith in the 1830s before producing his own range of spirits and helping to found the North British grain distillery, which is still running in the city today.

Now, better known as the name behind Scotland's most famous green ginger wine, enterprising drinks company Halewood International, which bought the company in 2007, have set out to revive the John Crabbie & Co. brand's reputation as a leading name in whisky and gin making.

Fortunate enough to be able to access 200 years of archives, one recipe book from 1837 provided a huge amount of inspiration with recipes for London Dry Gin, Old Tom Gin and fruit wines.

The distillery team were excited to discover that Crabbie was using some unusual ingredients and processes at the time, including adding sea salt to his base gin recipe. A few experiments of their own replicating the recipe yielded some fascinating results, with the sea salt helping to balance out the bitterness of some of the botanicals while also enhancing their flavours.

Having originally built an experimental distillery close to Crabbie's original premises, the team have now moved to a new purpose-built distillery on the city's Graham Street. Their product range has now expanded to include the original 1837 gin, followed up by their Old Tom, the Rugby Range (named for John Crabbie's sons who played for Scotland) and the Ginger and Wild Berry Gin Liqueurs.

Using their rich history of nearly two centuries of spirit making to produce not only these exciting gins but also a range of whiskies (including a soon to be launched single malt) the young diverse

BOTANICALS
Juniper, coriander, angelica, orris, liquorice root, grains of paradise, orange peel, Isle of Skye sea salt

PREFERRED SERVE
35ml Crabbie's Old Tom Gin and citrus tonic

PREFERRED GARNISH
Orange wheel

RECOMMENDED COCKTAIL
CRABBIE COLLINS

Ingredients
50ml Crabbie's Old Tom Gin
15ml lemon juice
25ml orange juice
15ml elderflower syrup
Soda water to top up

Method
1. Build over ice.
2. Garnish with orange peel and a lemon wedge.

team of John Crabbie & Co. distillers say they hope to be producing quality spirits at their Bonnington Distillery for at least another 200 years.

CRABBIE'S 1891 OLD TOM GIN

Named for the 1891 recipe it was inspired by, the intriguing thing about the Crabbie's Old Tom, is that they don't sweeten it using sugar like other examples of this style. In fact, much like the whisky industry, they use ex-sherry casks to impart extra flavour.

With this residual sweetness from the sherry, the producers also add orange peel, angelica root and coriander seed to the casks to steep, giving the gin a bit more colour and an extra hit of freshness.

A chance to highlight an under-explored category and a nice callback to the history of the drink itself, the Crabbie's team say that the Old Tom is a great way to bridge a gap and introduce customers to a sweeter style of gin – a step away from those traditional examples you will normally find in the supermarket.

LIND & LIME GIN

NOWHERE IS THE resurgence of craft Scottish gin more apparent than in the capital, where after over a century of absence, craft spirit distilleries are beginning to appear all over the city.

With established names such as Edinburgh Gin and Pickering's already forging a path for newer distilleries like South Loch, Crabbie's and Secret Garden, it's an exciting time for Edinburgh spirits.

However, one of the most thrilling examples comes in the form of the Lind and Lime, based in Leith, the region which made its reputation in centuries past as Scotland's gateway to the rest of the world.

Hoping to pay homage to Leith and the remarkable distilling, industrial, trading and scientific heritage it is known for, co-founders Paddy Fletcher and Ian Stirling hatched a plan that involved creating their own distilleries, whisky and of course a gin that would be a nod to Leith's rich connection to the spirits industry.

With such a remarkable range of gins now available out there, the pair decided they wanted to go back to the 'very definition of a classic London Dry Gin' – by creating a gin that was a delicate balance of juniper and lime with the gentle spice of pink peppercorns.

The inspiration for focusing on a recipe that used lime not only stemmed from the fruit's traditional connection to the spirit, most widely seen in the classic cocktail The Gimlet, but was also inspired by an unsung hero of Edinburgh's past.

James Lind, who was born in Edinburgh in 1716, was the surgeon on board the *HMS Salisbury* and is credited with conducting the first-ever clinical trial using lime to try to prevent scurvy, a disease encountered by sailors caused by a prolonged deficiency of vitamin C.

Thanks to his work, by the end of the eighteenth century, the Royal Navy was provisioning its ships with citrus fruit, leading to a remarkable health improvement in the sailors.

Currently, Paddy and Ian are engaged in some exciting projects in Leith, with

SOUTH-EAST SCOTLAND

SCOTTISH MARITIME

LIND & LIME GIN

PROVISIONED WITH LIME

44%vol   PRODUCT OF SCOTLAND   70cl ℮

24/26 Coburg Street, Edinburgh
lindandlime.com
Price £35 / Quantity 70cl / ABV 44%

BOTANICALS
Juniper, coriander, orris root, liquorice, angelica, lime, pink peppercorns

PREFERRED SERVE
50ml Lind & Lime over ice, topped with Bon Accord original tonic

PREFERRED GARNISH
Lime wedge

RECOMMENDED COCKTAIL
GIMLET

Ingredients
50ml Lind & Lime Gin
25ml freshly squeezed lime juice
Sugar syrup, to sweeten to taste

Method
1. Shake all the ingredients with ice.
2. Strain into a coupe glass.

construction underway on not one, but two spirits distilleries.

Firstly, a new home for Lind & Lime, situated on Leith Shore, an exciting new experience and event space opened in 2021.

The second being The Port of Leith Distillery, Scotland's first vertical whisky distillery which will be a major future visitor attraction for the capital in its own right and will feature amazing panoramic views of the Edinburgh skyline as well as creating a world-class single malt whisky. This is due to open in the summer of 2022 and will be the home for the Port of Leith Single Malt.

LIND & LIME GIN

As already mentioned, the founders wanted to create a traditional London Dry style gin, that has a classically crisp and refreshing taste. Made using seven botanicals including the key ingredients of lime and pink peppercorns, the resulting gin features leading notes of the juniper, backed up with fresh citrus and finishing with an oily pepperiness from the peppercorns.

The quality of the spirit is matched by the vessel it is delivered in, with a striking glass bottle reminiscent of a Victorian perfume bottle or medical flask. With tactile lines running vertically down the aquamarine tinted glass, it's one of the best-looking bottles out there.

Edinburgh Gin Distillery and Visitor Centre, Edinburgh and The Biscuit Factory Distillery, Leith
edinburghgin.com
Price £28 / Quantity 70cl / ABV 43%

31
EDINBURGH GIN

THE CAPITAL IS as rich in its heritage of distilling and brewing as it is in history and architecture so it might come as a surprise that up until the current crop of craft distilleries opened, there hadn't been a distillery exclusively producing gin in the city for nearly 150 years.

In the late eighteenth century, there were no less than eight licensed stills in Edinburgh. Perhaps even more incredibly, the legal operations were far outnumbered by the illicit stills at the time with the number suspected to be around 400 or so. The Port of Leith was also once a centre of trade for many types of spirit, including genever – gin's Dutch predecessor.

What's more, in 2018, the Scottish capital's residents were apparently drinking more gin per head than any other British city. And it's this demand for gin that has been successfully met by the launch of Edinburgh Gin, arguably one of the most successful Scottish brands outwith the big three – Caorunn, the Botanist and Hendrick's.

Founded by drinks industry veterans Alex and Jane Nicol in 2010, and now owned by family-run firm Ian Macleod Distillers, the brand took on distiller and graduate of Heriot-Watt's Brewing and Distilling course David Wilkinson in 2014 as part of a Knowledge Transfer Partnership with the university.

Edinburgh Gin made its home in the basement below the Rutland Hotel when it opened the new distillery and in-house gin bar Heads and Tales, where Flora and Caledonia – the brand's stills – are housed. Launched with its London Dry 'Classic' gin, Edinburgh Gin quickly followed up with the release of the Cannonball Gin, named for the city's rich naval heritage and the famous One o'clock Gun.

This 'Navy Strength' gin comes in at 100% proof (57.2% ABV) and features twice the juniper content of the classic Edinburgh Gin, with Szechuan peppercorns added to the recipe to create a spicy kick.

The popularity of Edinburgh Gin led

BOTANICALS
Juniper, orange peel, mulberries, milk thistle, coriander seeds, angelica root, orris root, liquorice root, cassia bark, lime peel, hazelnuts, pine buds, lemongrass, lavender

PREFERRED SERVE
In a tall glass with ice, using equal parts gin and good quality tonic water

PREFERRED GARNISH
A twist of orange peel

to the opening of a second distillery in an arts and fashion hub in what was once the old Crawford's biscuit factory in Leith. This second unit has the capacity to produce enough gin to create about 2.5 million G&Ts a year, and the brand is now in the process of building a new multi-million-pound distillery and visitor centre in the city's Old Town.

This exciting new development will allow the distiller to increase production capacity by over 200% while welcoming over 100,000 visitors through its doors annually, taking them on a one-of-a-kind

Edinburgh Gin Distillery

EDINBURGH GIN MARTINI

Ingredients
50ml Edinburgh Gin Classic
10ml Belsazar Dry vermouth
4 dashes of orange bitters
A twist of orange

Method
1. Add all the ingredients to a tall glass and stir.
2. Strain into a Martini glass.
3. Garnish with an orange twist.

journey of exploration of everything Edinburgh Gin.

Renowned for its flavoured gins and gin liqueurs, Edinburgh Gin is the UK's top-selling gin liqueur brand and now offers flavours such as Raspberry, Strawberry & Pink Pepper, and Pomegranate & Rose.

However, the most popular flavour from the Edinburgh-based distiller is Rhubarb & Ginger, which comes as a full-strength gin or a liqueur at 20% ABV. Since its launch in 2014, the sweet and fiery combination has helped to pioneer the flavoured trend which now occupies around 40% of the overall gin category.

EDINBURGH GIN

Described as a 'classic London Dry-style gin', this was the first expression to be released by the Edinburgh Gin brand. Created by hand in small batches at both the West End and Leith distilleries, Edinburgh Gin is made using a blend of 14 botanicals, including lavender, pine buds, mulberries and milk thistle. In designing a gin that tastes clean, zesty and juniper-forward, the EG team also used orange peel, lemongrass and lime peel to create fresh, uplifting citrus notes.

SOUTH-EAST SCOTLAND

HEIGHT OF ARROWS

32

BOTANICALS
Juniper, plus two modifiers – Isle of Skye sea salt and Edinburgh Honey Co beeswax

PREFERRED SERVE
This gin is made for sipping neat, just as you might enjoy a fine aged whisky, or with a simple straight tonic and ice

PREFERRED GARNISH
To retain the intended flavours, Holyrood recommends this gin without any flavoured-garnishes; but at the Holyrood Distillery it is served with an edible flower

Holyrood Distillery, Edinburgh
holyrooddistillery.co.uk
Price £34.99 / Quantity 70cl / ABV 43%

A HISTORIC OCCASION FOR the capital and the whisky industry as a whole, the opening of the Holyrood Distillery in 2019 was the first operational single malt distillery in central Edinburgh since The Edinburgh Distillery (aka Glen Sciennes) closed, almost a hundred years ago, in 1925.

This modern distillery, which is located on the edge of Holyrood Park close to popular Edinburgh Festival Fringe venues the Pleasance and George Square, reflects the vision of founder Rob Carpenter who wanted to create a new kind of distillery, one that would reflect the dynamism and diversity of present-day Edinburgh.

Rob, who founded the Canadian branch of the Scotch Malt Whisky Society together with his wife Kelly, has brought his vision to life in partnership with co-founder David Robertson, who has 25 years' experience in the industry.

Found in the heart of the capital within easy walking distance of the city's main attractions, including the National

Museum of Scotland, Edinburgh Castle and the Palace of Holyroodhouse, Holyrood is first and foremost a whisky distillery but through experimentation, they've created a gin which applies a whisky-maker's philosophy.

According to the team, they start building their whisky by its intention: what it is going to taste like, and where they want to go with it; and that is exactly what they have done with Height of Arrows.

Their desire was to produce a gin that is ideal for sipping, just as one might enjoy an aged whisky. They've done something very intriguing by completely paring back the recipe to very few ingredients – literally just juniper and two modifiers – Isle of Skye sea salt (for smoothness) and Edinburgh Honey Co beeswax (for texture).

Combining all three in their pilot still at their distillery, which sits under the shadow of Arthur's Seat itself, they've crafted a surprisingly complex gin, which they describe as a game-changer in such a saturated sector.

Drinks fans looking to find out how this exciting spirit is made, as well as their single malt whisky, can visit the distillery for a range of bespoke tours, guided tastings and a courtyard bar for an al fresco bar and street food experience.

RECOMMENDED COCKTAIL

ELVIS LIVES

Ingredients
25ml Height of Arrows Gin
25ml Campari
Schofferhofer Grapefruit Beer

Method
1. Fill a tall glass with ice.
2. Add the gin and Campari.
3. Stir.
4. Top up with Schofferhofer Grapefruit Beer Mix (approx 250ml).

SOUTH-EAST SCOTLAND

Pickering's Gin Distillery, Edinburgh
pickeringsgin.com
Price £32.95 / Quantity 70cl / ABV 42%

33
PICKERING'S

THE IDEA THAT would go on to form Pickering's Gin first surfaced in 2012 when Matthew Gammell and Marcus Pickering took possession of a hand-written scrap of paper, which dated back to 17 July 1947.

Given to them by a friend of Pickering's father, it detailed a traditional Bombay gin recipe made using nine botanicals. Based at Summerhall and having witnessed the successful launch of a brewery on the site, Pickering and Gammell decided to create a gin distillery on the part of the complex that formerly housed the dog kennels of the old Royal Dick School of Veterinary Studies, and perfect a re-creation of their Bombay recipe. In 2014 everything was ready and – after numerous trials to perfect their gin – a modern version of the original recipe was born. Named Pickering's after its founder, the flagship gin was soon followed by two more expressions, each marked out by the colours of their wax tops – red, orange and black.

The first, the Navy Strength, was created in 2014 in celebration of the brand becoming the official gin sponsor of the Royal Edinburgh Military Tattoo, while the second, the Original 1947, was made to the original recipe gifted to the pair on that scrap of paper.

The brand went on to create one of their most popular products in the winter of 2014 – with their award-winning gin baubles gaining notoriety as far away as the United States; the final December 2016 batch was a record-breaker with a run of 30,000 selling out in just 82 seconds. Sold in packs of six, the colourful baubles each contain a 50ml serving of Pickering's signature gin; over a million of the drinks-themed Christmas ornaments have now been sold.

Always innovative, they've recently revived their hugely popular Gin Jolly distillery tasting tours, launched several new flavoured gins and even expanded their Gin Bauble range with a new festive flavours pack that features a Brussel Sprout flavoured gin.

Fennel, anise, lemon, lime peel, cloves, juniper, coriander, cardamom and angelica

PREFERRED SERVE
In a glass with ice and tonic

PREFERRED GARNISH
Pink grapefruit

RECOMMENDED COCKTAIL
LAVENDER NEGRONI

Ingredients
25ml Pickering's Gin
25ml Campari
25ml lavender vermouth

Method
1. Add everything to an ice-filled rocks glass.
2. Stir gently for 45 to 60 seconds.
3. Garnish with a pink grapefruit twist.

NOTE
To create your own lavender-infused vermouth, mix dried lavender with bianco vermouth at a ratio of 10 to 1 vermouth to flowers. Leave it to macerate for at least 48 hours then strain out the flowers.

PICKERING'S GIN

Pickering's premium gin is made using nine distinct botanicals with fennel, anise, lemon, lime and cloves added to the traditional stalwarts of juniper, coriander, cardamom and angelica. The angelica was used as a substitute for the cinnamon to create a modern version of the original recipe that Matt inherited.

Described as 'a spectacularly smooth and flavoursome London Dry style gin', Pickering's is produced in Gertrude and Emily – the two 500-litre gin stills named after the founders' grandmothers – that feature a slow heating process which means Matt and the team can control the temperature to within one degree of accuracy. Each and every bottle of gin produced at the distillery is then labelled, corked and dipped in wax by hand before shipping; the distillery team say that this process of hand-finishing is still hugely important to them.

WILD GIN

FEW PRODUCERS CLAIM to have the botanical pedigree of renowned Scottish herbologist Hamish Martin, who founded and runs Edinburgh's Secret Herb Garden.

Nestled at the foot of the Pentland Hills close to the capital, the herb nursery was opened in 2014 when Hamish and his wife Liberty purchased a 7.5-acre site with the hopes of building their own commercial garden from the ground up.

Making the decision to use no chemicals whatsoever and encouraging wild indigenous plants to flourish, the garden is now home to over 600 naturally and sustainably grown varieties of herbs and plants, including a wide range of junipers, angelica and a dedicated apothecary rose garden, as well as a bee conservatory, a café, shop and, launched in October 2017, a gin distillery.

Having previously worked in the wine and spirits trade, Hamish started to experiment with herbs and spirits to demonstrate how remarkable nature is in naturally delivering both flavour and colour, with all of the plants used lovingly cultivated and hand-harvested to ensure the highest quality.

The botanicals are dried naturally at 37°C, distilled and then bottled at 39% ABV as Hamish's team believe this is best for the floral flavours to harmonise with the gin.

As of 3 May 2021, the business was rebranded as the Secret Garden Distillery and their gins became known as Secret Garden Gin. This was to forge closer links with the garden and accentuate their 'rooted in nature' values, which Hamish says remain at the heart of everything they do.

Crafted from 'seed to sip', the team have created a range of unique tasting gins full of natural flavours and aromas with no added sugars, artificial colourings, flavourings or additives, with the release of their original core gins in 2017 – the Apothecary Rose and Lavender & Echinacea gins – being followed up by the Lemon Verbena, Pink Elderflower and Jasmine, and Geranium and Mallow gins.

32A Old Pentland Road, Edinburgh
secretgardendistillery.co.uk
Price £35.95 / Quantity 50cl / ABV 39%

BOTANICALS
Juniper, coriander, elderberries, angelica, winter savory, bog myrtle, nettle, sweet cicely, Scots lovage, melissa, sweet woodruff, yarrow

PREFERRED SERVE
Wild is best served with a light tonic

PREFERRED GARNISH
A sprig of rosemary

RECOMMENDED COCKTAIL
GARDEN SPRING COLLINS

Ingredients
50ml Wild Gin
15ml lemon juice
20ml elderflower cordial
20ml rhubarb purée

Method
1. Shake and strain all the ingredients (except the cordial) over cubed ice into a highball glass.
2. Top with the cordial.
3. Garnish with a lemon wedge and a mint sprig.

In June 2018, the team followed up their gin launch with a UK first – a gin garden where visitors can observe the selection and picking of the botanicals, the drying process, right through to the distilling. The garden is packed with 1,500 juniper bushes – which take a few years to be ready – complemented by 80 herbs and floral varieties such as scented lemon verbena, geraniums, irises and roses.

WILD MADE WITH ORGANIC SPIRIT GIN
With wild plants considered to be a key passion of the Secret Garden team, an expedition with the RSPB to the Forsinard Flows Nature Reserve inspired the creation of a new gin in Wild. Bog myrtle, sweet cicely, nettle and six other indigenous botanicals are distilled with their organic spirit to create this unique gin.

One hundred per cent natural, this twist on a classic London Dry gin is naturally herbaceous and designed to 'reflect the wonders of nature', with a swing tag that is impregnated with seeds that can be planted directly into the ground to grow new plant life and a label made from 30% grass.

56 North, Edinburgh
southloch.com
Price £36.95 / Quantity 70cl / ABV 42.1%

SOUTH LOCH GIN

NOT CONTENT WITH a decade of helming one of the top gin destination bars in Scotland, Edinburgh's 56 North, and co-hosting his own gin podcast (with yours truly), James Sutherland also recently decided to launch his very own gin brand.

With over 400 gins behind the bar, James was able to draw on not only his own love of gin but also the knowledge of his staff and the preferences of his customers to decide which direction he wanted to go in.

Enlisting top distiller and Heriot-Watt Brewing and Distilling alumni Lindsay Blair in 2019, the pair, who combined have over three decades of experience in the drinks industry, set about experimenting with recipes and designing the branding for their new gins.

Using the historic name for the body of water in one of Edinburgh's most popular green spaces, the Meadows, which lies close to the bar, these exciting new spirits would pay homage to the southside's rich history of brewing and distilling.

BOTANICALS
Juniper, coriander seed, angelica root, fresh lemon peel and juice, fresh lime peel and juice, bitter orange peel, linder flower (lime flower), kaffir lime, cardamom, fresh ginger root, orris root

PREFERRED SERVE
South Loch Citrus and Lime Flower pairs beautifully with Fever-Tree Mediterranean tonic and a wedge of lemon

PREFERRED GARNISH
To accentuate the citrus notes in the gin, use either lemon or lime for a zesty, refreshing gin and tonic or a wedge of orange for a slightly sweeter note

The South Loch, as it was known up until 1621, originally provided much of the drinking water in the city and supplied water to some of the nearby breweries and distilleries.

SOUTHSIDE SOUR

Ingredients
35ml South Loch Citrus & Lime
 Flower
15ml St Germain elderflower liqueur
25ml fresh lemon juice
10ml sugar syrup
4 basil leaves muddled
2 dashes of celery bitters
1 egg white

Method
1. Muddle basil leaves and dry shake ingredients all together first without ice.
2. Add ice and shake hard and strain into a coupe glass.
3. Garnish with a basil leaf.

With a strong belief in traditional gins, James and Lindsay say their priority was to first and foremost create a range of gins that stood the test of time, but also with flavours that they both enjoyed drinking.

Focusing on a traditional juniper backbone, their gins feature vibrant and fresh botanicals in a bid to elevate them above a plain old London Dry resulting in two unique flavours, the Black Raspberry Old Tom and Citrus and Lime Flower gins.

Each of the gins is distilled in small batches at the 56 North bar, with gin fans able to visit and see the 50-litre copper still (hilariously named Ginny) in person.

Not only that, but they also host a range of excellent masterclasses, which include stunning cheeseboards and cocktails, in one of the most stylish settings in the city.

SOUTH LOCH CITRUS AND LIME FLOWER

Hand bottled by the team and coming in a striking lemon yellow ombré bottle, the label features nine geese in flight in a nod to the fact 56 North stands on what used to be a Guse Dub (goose pond).

Featuring a strong juniper background – something both James and Lindsay were very keen on as traditional gin fans – the Citrus and Lime Flower Gin combines typical botanicals with newer ingredients such as kaffir lime, bitter orange peel and linder flower (lime flower) to make a bright and flavoursome spirit that lives up to its name – delightful for enjoying in a beer garden on a warm summer's day.

THE GINS

CENTRAL SCOTLAND
PERTHSHIRE, STIRLING, FIFE, DUNDEE & ANGUS

Trossachs Distillery Limited, Callander
mcqueengin.co.uk
Price £23 / Quantity 70cl / ABV 37.5%

MCQUEEN

AS YOU HEAD north from Glasgow, you'll find the Perthshire town of Callander, which abuts the soaring peaks and gorgeous lochs of Loch Lomond and the Trossachs National Park.

The national park is home to a unique gin distillery, which was launched in June 2016 by chartered mechanical engineer Dale McQueen and his wife and business partner Vicky, a retired chef.

Tired of the daily grind, in the summer of 2015, the pair decided to leave their jobs and set up a distillery. As Dale tells it, he called his boss the next morning to resign then flew to Germany the week after to order the still, at which point the McQueens still had no building to put it in or so much as a licence to use it.

But from there, what would become the Trossachs Distillery slowly began to take shape. Self-taught in distilling, the McQueens turned their expertise in flavour and analytics to create 150 iterations of gin over a six-month period.

Then, taking in detailed feedback from around 3,000 members of the

BOTANICALS
Juniper, ghost chilli, Carolina reaper chilli, guajillo chilli, scorpion chilli, orange habanero chilli

PREFERRED SERVE
In a rocks glass with a Himalayan salt rim, over ice topped with ginger beer

PREFERRED GARNISH
A green or red chilli with a lime wedge

public, their core range began to emerge.

To their surprise, they found that they had the most success when diverging from the tried-and-tested formula of traditional flavours. Instead, they embarked on a path that would lead to them building a range of gins that included some 'world first flavour expressions' such as Sweet Citrus, Smokey Chilli, Spiced Chocolate Orange, Mocha and Chocolate Mint.

CENTRAL SCOTLAND

SMOKEY CHILLI NEGRONI

Ingredients
50ml McQueen Smokey Chilli Gin
50ml sweet vermouth
50ml Campari
Ice
Orange peel to garnish
Chilli slice to garnish

Method
1. Add the ingredients to a cocktail shaker and shake until cold to the touch.
2. Strain into an ice-filled glass.
3. Garnish with orange peel and a chilli slice.

The pair are keen to point out that, despite their distinctive profiles, this selection of expressions does not include flavoured gins created with post-distillation additions, but rather they are made in the traditional manner with everything happening during distillation and only water added post-process.

Over the next three years, their growing success saw an expansion and the First Minister arriving in 2019 to open their newly updated distillery and visitor centre.

Since then, they've not only updated and refreshed their core range, with new flavours including Citron, Five Chilli and the Black Cherry and Vanilla, the McQueen team have also unveiled what they believe are the 'world's coolest labels'.

At the start of 2021, they became the first drinks company in the UK to create a full augmented reality experience for their core range, with labels designed to create an immersive gin experience for customers.

By scanning the label via a dedicated app, each bottle showcases the hidden inner secrets of McQueen Gin with a unique full-length animation – with info on the local area and the exciting ingredients and flavours you'll find in each bottle.

FIVE CHILLI GIN

According to the McQueen website, Dale is a bit of a chilli fan so it's no surprise that a flavour profile such as this would feature in their range.

A balanced blend of that key juniper flavour with five of the hottest chillies on the planet; Ghost, Carolina Reaper, Guajillo, Scorpion and Orange Habanero, you'd expect it to be intense in its flavour with a lot of full-on heat, but surprisingly – or perhaps unsurprisingly given that it's made with the chilli added before distillation – it is actually a lot more subtle, with a warming, deep smoke reminiscent of mezcal.

BOË

THE BRAINCHILD OF Carlo Valente, Boë Gin is one of the Scottish scene's top success stories.

Distilled in the village of Throsk, near Stirling, they have grown from a small craft distiller to become one of the biggest independent gin producers in the UK.

Growing from quietly producing their gins, they've expanded to the point where they can afford to employ the likes of singer Paloma Faith to sell their wares.

The Boë Gin team say their brand is named for Dutch physician Franz de la Boë who – according to legend – created spirit with juniper berries while in search of a medicinal tonic. The team produce full-strength gins infused with the likes of passion fruit and their popular violet, as well as a range of gin liqueurs and their more traditional Scottish gin.

Boë Violet gin is instantly recognisable for its bright violet colouring. Created using their award-winning base spirit which is infused with violets post distilla-tion, it has a delicate floral flavour that's refreshing and delicious in equal measure.

BOTANICALS
Juniper, coriander, angelica, cardamom, ginger, almonds, orris root, cassia bark, liquorice root, cubeb and orange and lemon – with violets infused post-distillation

PREFERRED SERVE
Over ice with a premium tonic

PREFERRED GARNISH
Pink grapefruit twist

Unit 7c, Bandeath Industrial Estate, Stirling
boegin.com
Price £30 / Quantity 70cl / ABV 41.5%

The Old Smiddy, Stirling
stirlingdistillery.com
Price £34.99 / Quantity 50cl / ABV 55%

38
STIRLING GIN

THE HISTORIC CITY of Stirling is famous for its iconic landmarks such as its castle and, of course, the Wallace monument but, other than the nearby Deanston whisky distillery and the lost Craigend Distillery, it's been bereft of any form of spirit producer for centuries.

However, Cameron and June McCann, the husband-and-wife team behind the popular Stirling Gin brand, have changed that by finally bringing distilling back by opening a distillery and visitor centre in the shadow of Stirling Castle in 2019.

Now an excellent, fun-filled tourist destination, 'The Old Smiddy' distillery is housed in a building that dates back to 1888 and the site itself is rumoured to be where James V once kept his horses. The distillery offers gin tours and tastings, masterclasses, and even an on-site gin school.

Stirling Gin began its life in a tiny copper pot still named Jinty, in the McCanns' Bridge of Allan kitchen in 2015. It has since grown to become one of Scotland's most popular gins.

Handcrafted and small-batch, this distinctive gin derives its idiosyncratic herbal notes from a combination of basil and Stirlingshire nettles, with the early season nettles providing a slight sweetness of note, which enhances the taste and tickles the palate in a unique manner.

The nettles are hand-foraged from two historical locations that are intrinsically linked with the history of Scotland's medieval royal capital: the Minewood and the King's Park.

The Minewood is home to the medieval copper mine that was once used to produce currency for the Stewart kings of Scotland and the rolling green parkland of the King's Park was used for hunting by the Stewarts under the glowering watch of Stirling Castle's towering walls.

The McCanns have since released a lively range of gins including the Folklore Collection which features gins named after some local legends including the Green Lady, a mint and bramble liqueur named after the ghost who haunts Stirling Castle, and the Red Cap rasp-

Angelica root, juniper, lemon peel,
orange peel, basil and Stirlingshire
nettles

PREFERRED SERVE
Can be enjoyed straight, either neat or
over ice

PREFERRED GARNISH
Basil leaf and orange peel

RECOMMENDED COCKTAIL
STIRLINGTINI

Ingredients
25cl Battle Strength Gin
25cl Red Cap raspberry liqueur
100cl pineapple juice

Method
1. Place all ingredients in a cocktail
shaker with ice.
2. Shake and serve in martini glass.

berry liqueur, which gets its name from a
mythological and murderous goblin.

Ever inventive with their gin range,
they've also added a vibrant pink gin,
which is already proving popular, and
the summer sipper Tropical Triumph,
which is produced using the delicious
tropical flavours of mango, passion fruit,
pineapple and lime. Keep an eye out
for their Sons of Scotland Independent
whisky bottling range, as it's a precursor
for their own single malt whisky, which is
also on its way.

BATTLE STRENGTH GIN

Telling the story of Stirling's past, in this
case the famous battle of Stirling Bridge,
the Battle Strength – which Cameron
jokes is so named because the city has
never really had strong links to the navy
but has had plenty of battles – comes in
at a lip smacking 55%.

Another nod to the city's heritage
(Robert the Bruce died in his 55th year)
this incisive gin is the big brother to the
nettle-infused original. Packaged in a
bottle which has a hand-dipped wax seal
and an illustration of the famous battle
by Scottish artist Ritchie Collins, this gin
is designed by Cameron to fill that gap
between gin and whisky.

KILTY

STIRLINGSHIRE IS A stunning area that sits at the heart of the country and is home to some exciting gins.

One of the newest, Kilty Gin, was founded by entrepreneur and dentist Richard Thurlow-Begg. Drawing inspiration from his grandfather, a wine maker, Richard developed a passion for distilling.

Infatuated with the idea of starting his own distillery, after just two years of research, and hundreds of recipe trials, the first batch of Kilty Gin started to flow off his Dunblane-based still in spring 2019.

As for the intriguing name, Richard references the happy memories he had from travelling the world as a student and the great reception his kilt got and the wish to capture that feeling with his gin.

Produced on a 250-litre still, this London Dry style gin features the signature botanical of rowan berries, which Richard picks in late summer and early autumn and freezes. Traditionally thought to ward off evil spirits, the berries thankfully also add a uniquely sweet but crisp flavour to great Scottish spirits too.

BOTANICALS
Juniper berries, coriander seeds, rowan berries, angelica root, pink peppercorns, liquorice root, oranges, cardamom pods, lemon peel, heather tips

PREFERRED SERVE
Served with a classic Indian tonic and 3-4 cubes of ice

PREFERRED GARNISH
A wedge of pink grapefruit for summer and a sweeter palate. Or a strip of fresh ginger for a more savory palate and a winter month alternative

Dunblane, Stirlingshire
kiltygin.com
Price £36 / Quantity 70cl / ABV 42%

CENTRAL SCOTLAND

Darnley's Distillery, Fife
darnleysgin.com
Price £34 / Quantity 70cl / ABV 42.7%

DARNLEY'S

AFTER BEING CONTRACT distilled in London, Darnley's Gin – previously known as Darnley's View Gin – enjoyed something of a homecoming in 2017. Returning to the East Neuk of Fife – the traditional home of owners, the Wemyss family – it is now housed in a small farm cottage found on the grounds of its sister distillery, Kingsbarns, which produces the stunning single malt whisky, Dream to Dram.

Born of the Wemyss family's passion for spirits, Darnley's is the result of their background in whisky and wine, but also their fascination with the endless combination of botanicals you can use in gin making and the impact this has on flavour.

The name 'Darnley' is a celebration of the moment that Mary Queen of Scots first spied her husband-to-be Lord Darnley at Wemyss Castle in 1565, the ancestral family home.

The gin underwent a rebranding in 2017, with the word 'view' being dropped from the label. The packaging was also updated to reflect a more modern approach, with sketchings of the botanicals added and the labelling pared back.

Darnley's' core range now features three main gins – Original, Spiced, and Spiced Navy Strength – and a limited edition range dubbed the Cottage series, most of which are now made in their 350-litre Italian copper still, named Dorothy after two of their regular visitors.

Fife local Scott Gowans is the gin distiller heading up production at this

BOTANICALS
Juniper, cinnamon, cassia, angelica, coriander, nutmeg, clove, cumin, ginger and grains of paradise

PREFERRED SERVE
In a tall glass with ice and ginger ale

PREFERRED GARNISH
A slice of orange

DARNLEY'S RED SNAPPER

Ingredients
50ml Darnley's Spiced Gin
75ml tomato juice
10ml fresh orange juice
Juice of half a lime
3 dashes of Tabasco sauce (to taste)
2 tsps Worcestershire Sauce
2 grinds of cracked black pepper
2 tsps fino sherry
Celery salt
1 stick of celery to garnish

Method
1. Run a lime wedge around the rim of the glass and then dip the wet rim into the celery salt.
2. Fill the glass with cubed ice then pour in the gin, tomato, lime and orange juices and the Tabasco and Worcestershire sauces.
3. Float the sherry on top and grind black pepper over the top too.
4. Garnish with a celery stick.

fascinating east coast distillery and the company recently launched their very own gin school, where visitors can have a go at being a distiller for the day by choosing their own botanicals, running the copper still and even labelling and bottling their own gin to take home.

Darnley won the Scottish Gin Destination of the Year award in 2020.

DARNLEY'S SPICED GIN

Darnley's Gin is made in a traditional pot still using the London Dry method, and though the original is a great example of the style, the spiced gin – launched in 2012 – intrigues with its own particular *je ne sais quoi.*

While its predecessor was inspired by the elderflower that grows wild around the Wemyss family estate in Scotland, the spiced version is influenced by a faraway range of global flavours. Featuring the usual players such as juniper, angelica and coriander, the spiced gin also contains the tastes of South Asia and West Africa.

Ingredients such as nutmeg, clove and grains of paradise give it a wonderfully rich flavour that sets it apart in an enticingly distinctive way from many of its competitors.

MISS G'S GORGEOUS SCOTTISH GIN

WHAT COULD BE be more exciting for any gin lover than realising the dream of creating your own gin? Gail Sneddon had been toying with the idea of doing just that and had been working on the design with her husband Stewart, when he tragically passed away in 2017. She almost left it there, but her friend Steven encouraged her to continue making her gin. The entrepreneur launched Miss G's Gorgeous Scottish Gin, using Stewart's designs, in May 2020.

Working with contract distiller Stuart Ingram at House of Elrick and spirits consultant David Clutton, Gail explained that Gorgeous featured the botanicals and flavours she loves. Featuring a beautiful square bottle complete with Vinolok glass stopper, Gorgeous Scottish Gin is designed to stand out, replete with a love heart that is a nod to Stewart.

Made using nine carefully chosen botanicals, it features a citrus base from the lemon verbena and pink grapefruit along with a chorus of herbal, spice and sweet notes.

BOTANICALS
Nine botanicals including juniper, angelica, liquorice and pink grapefruit

PREFERRED SERVE
Over ice with a premium botanical

PREFERRED GARNISH
Slice of grapefruit and a sprig of rosemary

Newport on Tay, Fife
gorgeousgin.com
Price £40 / Quantity 70cl / ABV 42%

CENTRAL SCOTLAND

Tayport Distillery, Fife
tayportdistillery.com
Price £25 / Quantity 50cl / ABV 40%

WILD ROSE GIN

PULLING ON THE heritage of moonshiners in her home state of Ohio, USA, and inspired by the stories of the growing number of craft distillers in her adopted country of Scotland, Kecia McDougall left her full-time job and set out to make Scotland's first eau de vie, a clear spirit made using fruits, in 2017.

Setting up their family-run distillery on the outskirts of the Tentsmuir National Nature Reserve on the east side of Tayport, and born out of a love for Scottish produce, Kecia and her daughter Mary wanted to showcase the incredible soft fruits growing on their doorstep in Fife.

Following the creation of a bespoke gin in support of a local charity, their loyal customer base, who they regularly engage with at farmers markets, kept asking if they'd ever create another.

Deciding it was a good idea they set out to develop a spirit that was in keeping with their hallmark of highlighting locally sourced ingredients.

Enlisting the help of an expert chef from the nearby Newport restaurant, the pair went foraging around the surrounding area and discovered the exciting wild rose. No other distilleries were using this unique botanical, growing on the banks of the Tay, and the mother-daughter team decided it would be the ideal botanical to build a gin around.

Showcasing the idea that the distillery has so much more to offer, they decided to add a Barley Vodka to their core range. Developed during the lockdown period using the skills they had built up over three years of making their eau de vie and fruit liqueurs, it has already gained awards and been recognised as one of the world's best by Forbes.

With their new gin being well received, and Kecia's son joining the team as an apprentice distiller, the business is going from strength to strength. In addition, while the Wild Rose Gin is considered to be their accessible, summertime gin, their follow up, the Scots Pine is a much punchier spirit with

Juniper, coriander, lemon peel, wild rose, angelica, almond and bay leaf

PREFERRED SERVE
A classic light tonic over lots of ice

PREFERRED GARNISH
Lemon or orange. Orange for more sweetness

RECOMMENDED COCKTAIL
LEMON & CARDAMON GIMLET

Ingredients
37.5ml lemon and cardamom simple syrup
5ml Wild Rose Gin
Dehydrated lemon wheel

Method
1. Add the gin and simple syrup to a shaker with ice and shake until well-chilled.
2. Strain into a chilled cocktail glass.
3. Garnish with a dehydrated lemon wheel.

heavier notes of warm winter spices.

Those curious to see how this small family-run distillery creates their range of award-winning gins should pay them a visit – the Tayport Distillery team offers excellent guided tours.

WILD ROSE GIN

This hugely accessible and lightly floral gin is created with a velvety smooth base spirit made from local grain, and produced on their 500-litre still, which is named Mamó, after the Gaelic for grandmother. The moniker is a loving nod to their much-missed grandmother who sadly passed away from MND; the business is also a firm supporter of their local MND charity.

Offering a little taste of Scottish summer, it's a really great introductory gin and is perfect for those who aren't seeking those large snappier notes, with the enticing juniper base followed by the citrus from the lemon peel before those lovely subtle floral notes from the Wild Rose shine through on the finish.

EDEN MILL

BUOYED BY THE success of their beer and gin, and with a strong desire to place sustainability at the heart of what they do, St Andrews' own Eden Mill (which can be found close to the banks of the River Eden) has started development on a £10-million venture to create Scotland's first carbon-neutral single malt whisky distillery at Guardbridge on the site of the old Seggie Distillery. This will see the site of a former paper mill transformed into a state-of-the-art brewery and distillery.

Eden Mill are leading the charge in gin tourism, first with their unique Blendworks experiences, then with their innovative (and great value) virtual tastings which eventually eclipsed the number of actual guests expected at the distillery. They aim to have over 50,000 visitors a year walk through the doors of their new visitor centre, which overlooks stunning views of the Eden Estuary, right up to St Andrews' famous Old Course.

Not one for resting on his laurels, founder Paul Miller also released the

distillery's first single malt whisky in 2018.

This marked the end of the journey that saw them lay down their first casks in 2015; it also cemented their place as Scotland's first true single-site brewery and distillery, with the country's triumvirate of drinks – whisky, beer and gin – set to all be produced on the one site.

Starting life as a craft brewer in 2012, the company expanded into spirits in late 2014 with the launch of its original gin

CENTRAL SCOTLAND

Eden Mill Distillery, St Andrews
edenmill.com
Price £25 / Quantity 70cl / ABV 40%

created using an intriguing botanical – sea buckthorn berries sourced in Fife – it won Gin of the Year at the Scottish Gin Awards in 2018.

A hop gin, which alluded to their connection between brewing and gin making, and the Oak gin, which is made using oak chips to mellow out the flavour, were then developed by the forward-thinking distillery team.

Other seasonal specials followed, including a nod to St Andrews and its links to golf with a gin made using hickory wood, traditionally used to make golf clubs.

They've also recently updated their entire range, as well as curating a selection of premium RTD cocktails, adding a zero alcohol G&T 'Eden Nil' in 2018, and exciting fans with the launch of a collaboration with international chef and restaurateur Gordon Ramsay.

But it's Eden Mill's Love Gin, the brand's famous pink gin, which is the focus here.

EDEN MILL LOVE GIN

Eden Mill made the move away from their ceramic bottles to more sustainable crystal-cut bottles, which use 18% less glass than other industry-standard bottles, as part of their continuing drive towards carbon neutrality.

This means the Love Gin now comes in a pretty (and tactile) bottle which

shows off its stunning colour, it really comes into its own on a warm summer's day when its deliciously fruity flavour adds just the right level of refreshment to a mixed drink.

Featuring rose petals, marshmallow root, goji berries, rhubarb root and hibiscus flowers, as well as some of the more classic gin botanicals, the Love Gin has a floral taste with warm berry notes. Its lovely pink colour means the gin appeals to the eyes almost as much as to the nose and taste buds.

GARDEN SHED GIN

BOTANICALS
Juniper, blackberries, dandelion root, bay leaves, liquorice root, elderberries, grains of paradise and lavender

PREFERRED SERVE
In a glass with plenty of ice and Fever-Tree Original Tonic

PREFERRED GARNISH
Blackberries and a sprig of rosemary

Eden Mill Distillery, St Andrews
thegardensheddrinksco.com
Price £37.50 / Quantity 70cl / ABV 45%

A S IF TO highlight just how accessible the gin scene now is, the Garden Shed Drinks Company was set up by two Glaswegian couples in, well, their garden shed.

Scottish international rugby union player Ruaridh Jackson and teammate Ryan Grant, along with their wives Kirstin and Maxine, hit upon the idea to create their own gin using botanicals sourced in their back garden in Scotstoun, alongside one or two more traditional ones.

After a few experiments with different recipes, they settled upon one they found appealing and, enlisting the knowledge of the Eden Mill Distillery team – a partnership bred of the St Andrews brand's sponsorship of Jackson and Grant's Glasgow Warriors team – they developed their Garden Shed Gin.

Made using 14 selected botanicals, including brambles and dandelion root (sourced at home) along with juniper, grains of paradise and lavender, the gin is bottled in rustic packaging to reflect the nature of its origins.

Kirstin and Maxine designed the bottle which has a vintage feel furnished with wooden tags and twine.

Now cuckoo-distilled at Strathearn Distillery, they've recently added a Côte-Rôtie Aged Gin – which uses an old wine barrique to impart incredibly rich flavours – and the Bramble Peach, which is distilled with a delicate balance of botanicals including fresh blackberries, and flavoured post-distillation with a combination of blackberry (bramble) and peach.

The provenance of their product and the sustainability of the ingredients they use are hugely important to the quartet. Some of the profits from the sale of each of their gins are donated to important charities such as Trees4Scotland and the Bumblebee Conservation Trust.

With its sweet and citrus notes, this is a well-balanced gin that's dangerously drinkable, even on its own.

RECOMMENDED COCKTAIL
GRUNTING GROWLER

Ingredients
50ml Garden Shed Gin
1/3 of a can of ShinDigger Mango Unchained IPA (or other sweet IPA)
A few dashes of passionfruit juice
A few dashes of yuzu juice

Method
1. Pour the gin into a tall glass filled with the IPA.
2. Add the dashes of passion fruit and yuzu juices.

CRAG & TAIL

BOTANICALS
Watermelon, elderflower, orange, lemon, kaffir lime leaf, coriander, angelica, liquorice, orris root, cubeb, cassia, juniper

PREFERRED SERVE
Premium tonic (Fever-Tree Mediterranean, Original or Light)

PREFERRED GARNISH
Wedge of Lime

Huffmans Limited, Perth
cragandtail.co.uk
Price £36 / Quantity 70cl / ABV 41%

FOUNDERS GREGOR MACLEAN and Ramone Robertson's original connection to the Scottish gin market was as facilitators, with their company Huffmans set up in 2016 to provide a route to market for all of the emerging Scottish spirits coming onto the scene.

However, after a year or so of being embedded in the market, the pair decided it might be an idea for them to have their own flagship gin and Crag and Tail was born.

Linking up with expert distiller Lewis Scothern who was then based at one of their customers, Ogilvy – the nearest distillery to their hometown of Forfar – they set about their recipe research.

Quickly learning what flavours they liked and didn't like, they also wanted to use a distinctive signature botanical that would be fresh and crisp, which is where the fresh watermelon came in.

Delighted with the resulting recipe, Gregor explained that they then sat on 300 bottles for around a year, before they hit upon a design and name they

TOM COLLINS TWIST

Ingredients
50ml gin
25ml St Germain Elderflower
25ml lemon juice
125ml soda
Slice of lemon

Method
Build the cocktail in a glass 1 by 1 in the following order:

1. Ice to start
2. Gin
3. Lemon Juice
4. St Germains
5. Top up with soda
6. Garnish with a slice of lemon

both liked, with Crag and Tail referencing a major geological feature of which Ailsa Craig and Dundee's Law Hill are prime examples.

The duo have also launched 'Crag Culture'; spearheaded by Ramone, it's a new campaign to incentivise as many people as they can to get involved in activities in the great outdoors.

CRAG AND TAIL GIN

Designed to be a laid-back all-rounder, this is the kind of crisp and refreshing London Dry gin that is complex enough that each of the key flavour elements are there in perfect balance, with the floral notes, hints of spice and citrus all working in harmony.

As Gregor points out, it's a go-to gin that you can easily go through a bottle of at the weekend with your friends.

CENTRAL SCOTLAND

Strathearn Distillery, Perth,
teasmithgin.com
Price £38.95 / Quantity 70cl / ABV 43%

TEASMITH

I N THE CURRENT boom for Scottish gins, it can be hard to find a USP that allows you to really stand out from the crowd.

However, with Aberdeenshire-based company The Teasmith you have the first gin distilled in Scotland (currently produced by Perthshire's Strathearn) that uses hand-picked tea as one of its key botanicals. Sourced from the Amba Estate in Sri Lanka, the special Ceylon tea was chosen as a key botanical thanks to the citrus notes and distinctive minty sweetness it provides when distilled.

The Teasmith also reflects the area's rich and largely unknown heritage links with the tea trade, with Aberdeen harbour being famous for its shipbuilding of old, with several notable tea clippers built there. However, it's the story of James Taylor from Auchenblae, near Laurencekirk, who is the gin's true inspiration. As a young man in 1815, he left Aberdeenshire and headed for Sri Lanka, which at the time was a coffee-growing nation called British Ceylon.

BOTANICALS
Juniper, coriander seed, Ceylon tea, dried orange peel, rose petals, orris root, angelica root, liquorice root, honeyberry, grains of paradise and calamus root

PREFERRED SERVE
Copious amounts of ice and a premium tonic. Equally the gin is very smooth neat – the sweetness of the tea as it rounds the profile means you can enjoy it simply over ice on its own or with a mint leaf or two

PREFERRED GARNISH
Fresh mint leaves – the Teasmith team find these really accentuate the mint flavour tones that the tea provides when distilled. If you don't have mint to hand, a simple lemon peel twist also works really nicely

The island's crop was suffering extensively from crop disease. Taylor had heard about the success of tea growing

CLASSIC SOUTHSIDE

The simplicity of the ingredients and the predominant citrus and mint flavours, with their slight sweetness, mean this cocktail marries exactly with the spirit's flavour profile.

Ingredients
60ml Teasmith Gin
30ml fresh lime juice
15ml sugar syrup
8 mint leaves
Champagne
1 mint sprig to garnish

Method
1. Add all the ingredients to a cocktail shaker with plenty of ice.
2. Shake vigorously and fine-strain into a long Champagne flute.
3. Top up with a splash of Champagne for a touch of fizz and garnish with a small sprig of mint.

in India and decided to create the island's first commercial tea plantation. It was so successful that the island soon switched to tea growing.

Taylor set the wheels in motion that have helped turn Sri Lanka into one of the most significant tea-growing regions in the world. To this day, Taylor is revered as the 'Father of Ceylon Tea'. Working with tea consultant Beverly Wainwright to source the hand-picked and hand-rolled Sri Lankan tea, Nick and Emma Smalley from Udny Green, began formulating the idea for a gin in May 2015.

After 16 months of experimenting and forming the brand, they launched Teasmith Gin in December 2016 – the company has gone from strength to strength ever since, recently exporting to the US with a major retailer.

They have also just supported Balmoral Castle and Estate with the launch of Ballochbuie Gin which features ingredients foraged from the estate. The Teasmith has a very classic flavour profile with a strong juniper profile backed up by citrus notes and a refreshingly sweet finish.

BADVO

ONE OF THE youngest entrepreneurs on the Scottish scene, Helen Stewart founded her own gin distillery on her family farm aged just 22. Having fallen in love with distilling working for a local whisky distillery, she discovered a family heritage in distilling, giving her the thirst to follow in their footsteps.

While working towards her degree, Helen renovated a building on their Badvo hill farm, outside Pitlochry, and sourced funding for her new dream. Now the base for Helen's stills – named Archie and Mags after family friends – she uses botanicals foraged entirely from the farm to create gins including her signature Badvo Gin, the follow-up 1451 Gin and her Badvo Sloe Gin Liqueur.

Helen uses foraged botanicals including juniper, nettles, wild mint, honeysuckle and meadowsweet to create this complex gin which features sweet, herbaceous notes that are followed up by a refreshing mint blast on the finish. Each batch is hand-bottled by Helen and her mum.

BOTANICALS
Scottish juniper, nettles, rowan berries, green apple, honeysuckle, wild mint

PREFERRED SERVE
A standard tonic with a little ice

PREFERRED GARNISH
Green apple

Badvo Distillery, Pitlochry
badvo.com
Price £39.95 / Quantity 70cl / ABV 45%

CENTRAL SCOTLAND

133

PERSIE

BOTANICALS
Juniper, freshly zested limes and oranges

PREFERRED SERVE
Serve as a G&T with full-fat or elderflower tonic

PREFERRED GARNISH
Lime and torn mint

Persie Distillery, Bridge of Cally
persiedistillery .com
Price £29 / Quantity 50cl / ABV 42%

LYING JUST SOUTH of the borders of the Cairngorms National Park in the quiet setting of Glenshee, is Persie Distillery. What began as a simple gin tasting on World Gin Day in June 2014 has grown to become a hugely successful distillery with not one but six core gins.

The idea for the distillery sprung forth while founder and Institute of Brewing and Distilling graduate Simon Fairclough was touring the country with Gin Club Scotland, helping not just Scots but also gin fans around the UK to find their 'perfect snifter'.

Taking up residence in the former Persie Hotel at the foot of Glenshee in Perthshire, Simon set about turning the discoveries he had made about gin understandings and preferences into a physical product in a bid to give consumers what they want.

In May 2016, the Persie team launched three original gins, each designed with the Scottish gin palate in mind and reflecting the three favourite gin profiles of gin drinkers in Scotland –

HUGO

Ingredients
2 measures Persie Zesty Citrus Gin
1 measure elderflower cordial
A few leaves of fresh mint
Champagne or Prosecco
Mint or elderflower to garnish

Method
1. Put the mint, elderflower cordial and gin into a cocktail shaker.
2. Add ice and give it a good shake.
3. Double-strain into a Champagne flute, filling to about halfway up.
4. Top up with Champagne or Prosecco.
5. Garnish with some mint leaves or elderflowers.

still. They are made using water sourced directly from the local hills surrounding Glenshee, a place traditionally – and poetically – known as the glen of fairies.

Persie is unique in that it approaches gin from not only the angle of taste and production styles but also, and more significantly for them, of aromas. Simon explains that the customer feedback they received showed that a great gin needs more than just a great taste: it needs to stimulate an individual's sense of smell.

PERSIE ZESTY CITRUS GIN

The gins that the Persie team has created are made with carefully chosen botanicals to evoke an emotive and comforting scent. Zesty Citrus is described as sharply fruity, chock-full of limes and oranges; Herby & Aromatic is dry and savoury, with rosemary and basil; and Sweet & Nutty Old Tom is full-bodied and creamy, with fresh vanilla pods, almonds and a hint of root ginger.

The Zesty Citrus is described as a 'burst of limes and blood oranges on the nose, with a sharp citrus cut-through and long, zingy finish'. It is the perfect gin for those looking for fruitier flavours both on the nose and on the palate.

It takes around 200 fresh limes and oranges to create each small batch of roughly 300 bottles, which means that this gloriously summery gin offers exactly what it says on the label.

fruity, savoury and sweet – just as Simon learned during his travels.

Then, two years later, Persie Distillery joined forces with PADS (Perthshire Abandoned Dogs Society) and released a family of 'dog gins' to raise money for dog rescue. These expressions have been carefully designed to reflect dog breeds, with the mellow and classic Labrador Gin; wild and spicy Spaniel Gin; and sweet but sharp Dachshund Lime Gin Liqueur.

All the gins are handmade in small batches in a bespoke 230-litre copper pot

CENTRAL SCOTLAND

THE GAEL

BOTANICALS
Juniper, cardamom, angelica, orris root, coriander seeds, orange peel, lemon peel, liquorice, cassia bark, heather

PREFERRED SERVE
Fever-Tree Mediterranean light tonic and a slice of orange

PREFERRED GARNISH
Orange and rosemary

The Gael Spirits Company Ltd,
Wester Essendy
thegael.co.uk
Price £40 / Quantity 70cl / ABV 40%

WHAT HAPPENS WHEN you combine a love of music with a love of gin? Well, that's the question Nigel and Beverley Large hoped to answer when they teamed up with friends Jamie MacLean and Tanya Brown. The result was a spirit inspired by 'The Gael', a fiddle tune that was originally written by Dougie MacLean in 1988 and featured in the movie *The Last of The Mohicans*. Working with experienced distiller Rickie Christie at Strathleven Distillers in Dumbarton they launched their new gin in 2018.

The striking bottle features the musical score for the song printed on the inside of the label, their music-themed gin is made using malted barley to emphasise the link with traditional Scottish spirits.

Distilled five times in a bespoke hybrid still for added smoothness, ten botanicals including Scottish heather are added to produce a profile that features a citrus aroma with hints of floral notes. The creaminess of the barley spirit works well with the juniper and slight touches of spice to create a smooth warm finish.

VERDANT

KNOWN AS THE city of jute, jam and journalism, Dundee is the birthplace of Desperate Dan, marmalade and the famous Antarctic expedition ship, the *RRS Discovery*.

Often overshadowed by Scotland's other cities, Dundee has undergone a cultural revolution in recent years and, spearheaded by the 2018 opening of the V&A museum, has found itself listed in numerous travel guides as one of Europe's up-and-coming destinations.

The city is also home to the winner of the first-ever 'Best Gin in Scotland' award. Verdant Dry Gin is the first release from Verdant Spirits, a company founded in April 2017 by Andrew Mackenzie – it also happens to be the first legal distillery in the city of discovery in nearly 200 years.

Mixing tradition with modern techniques, the distillery is based in a former engine house of an old mill.

Formula One industry veteran Mackenzie went back to university in Dundee to study a master's degree in Food and Drink Innovation and

BOTANICALS
Juniper, coriander seed, lemon peel, bitter orange, cassia bark, orris root, green cardamom, angelica, liquorice and grains of paradise

PREFERRED SERVE
Either neat or with a premium tonic in a big glass, lots of ice and a wedge of lime or a twist of orange peel

PREFERRED GARNISH
A wedge of lime or a slice of orange peel – but be creative. For instance, renowned Scottish chef Tom Kitchin serves Verdant with a cinnamon stick and some juniper

Verdant Spirits Ltd, Dundee
verdantspirits.co.uk
Price £34.95 / Quantity 70cl / ABV 43%

CENTRAL SCOTLAND

Technology, where he was inspired by a lecture by the current chair of the Scottish Craft Distillers Association, Alan Wolstenholme, on the spirits business.

Taking a keen interest in what he perceived as a gap in the market for quality spirits for the home cocktail market, Mackenzie set about creating the first of a new range of spirits.

Distilled in small batches on their 500-litre still – named Little Eddie – Verdant Dry Gin is a nod to not only the classic style of gin but also the city's rich connection with the global trade routes of the past. Mackenzie chose each of this gin's ten botanicals specifically to give an overall flavour profile that's balanced and nuanced, allowing each ingredient their place in the final taste

Verdant fought off competition from 16 other contenders to take gold medal in the London Dry category at the Scottish Gin Awards 2017 before going on to win 'Best Overall Gin'. Buoyed by such success, Mackenzie and his team followed up this success by winning the tender to become the sole supplier of the special House of Commons Gin, in 2019.

A classic London Dry gin, Verdant is designed to be as flavoursome as possible with rich top notes of juniper being followed by hits of citrus and an undercurrent of spice.

It's complex, full of flavour but still fresh-tasting, and is made with mixology in mind.

ELDERFLOWER COSMOPOLITAN

The cosmopolitan sprang to life in the 1980s but can trace its family tree right back to the gin gimlet, and even further to the classic 'Daisy'. Here, the addition of the elderflower liqueur adds a wonderful touch of sophistication to this perennial cocktail that still lets the full flavour of the Verdant shine through.

Ingredients
25ml Verdant Dry Gin
15ml triple sec
20ml elderflower liqueur
5ml fresh lime juice
35ml cranberry juice
1 slice of blood orange, to garnish

Method
1. Place all the ingredients into a cocktail shaker and fill with ice.
2. Shake well and then strain into a coupe or martini glass.
3. Garnish with a slice of blood orange.

Produced using ten botanicals, including bitter orange, lemon peel and liquorice, this gin is perfect for those who enjoy that traditional, juniper-led style.

BLACK THISTLE

BRECHIN-BASED SPIRITS producers Black Thistle Distillers first arrived on the scene in 2018.

Specialising in a range of not only unique gins, but rum and vodka too they take full advantage of the region's reputation for producing some of the finest fruit, vegetables and, of course, botanicals in the country.

Beginning life in their recipe development lab, their spirits are allowed to rest for a week before they are tweaked and refined to create the final spirits that are then scaled up on their bespoke German-made 800-litre batch stills.

Utilising a library of 53 botanicals, they've created everything from the traditional London Dry gin, to the more exotic in the form of their shimmering mist gins.

The recipe for this London Dry fuses traditional botanicals such as green juniper and warm, aromatic coriander, with their own hand-picked blend of fragrant orris root, caraway seeds, wormwood and the wild thistle which grows in abundance around their distillery.

BOTANICALS
Juniper, coriander, orris root, caraway seeds, wormwood

PREFERRED SERVE
Over ice with a premium tonic

PREFERRED GARNISH
A twist of citrus – such as lemon or orange

Unit 8, Brechin Business Centre, Brechin
blackthistledistilleries.com
Price £37 / Quantity 70cl / ABV 41%

CENTRAL SCOTLAND

139

Arbikie Highland Estate Distillery, Arbroath
arbikie.com
Price £43 / Quantity 70cl / ABV 43%

NÀDAR GIN

NESTLED AT THE foot of the Angus hills, Arbikie is a working farm, which launched a spirits distillery on their grounds in 2014. It was the brainchild of the estate's fourth-generation farmers, founders David, Iain, John and Sandy Stirling.

The brothers decided they wanted to revive 'field to bottle' distilling, taking produce grown on the farm and using it to create a diverse range of spirits that began with the creation of what is considered to be Scotland's first commercially available potato vodka.

They went on to produce several innovative vodkas, but it wasn't long before they moved into the category that would cement their reputation as one of Scotland's most fascinating drinks producers.

To get the project up and running, the brothers hired Heriot-Watt alumni Kirsty Black and Christian Perez-Solar as master distiller and production manager to oversee the creation and launch of the new distillery.

BOTANICALS
Lemongrass, citrus leaf

PREFERRED SERVE
In a tall glass filled with ice. Add a generous measure of Nàdar Gin, top up with a quality light tonic

PREFERRED GARNISH
Ginger slice

Kirsty, who originally enrolled in the master's degree in Brewing and Distilling at Edinburgh University in a desire to learn more about brewing, is one of the pioneering female master distillers in Scottish gin.

The duo played a large part in the creation of Arbikie's initial flagship expression (now named after its creator, Kirsty), which was designed to embody the elements of the sea, rock and land that surround the 2,000-acre estate. An

NÀDAR GIMLET

Ingredients
50ml Nàdar Gin
25ml lime juice
15ml sugar syrup

Method
1. Chill a cocktail glass in the fridge.
2. Add the ingredients to a shaker with ice then stir until the outside of the shaker feels cold.
3. Strain the mixture into your chilled glass.

named after the Stirlings' father, which is made from fresh honey from their farm's bees.

With their keen interest in taking the gin from 'field to bottle', Arbikie are proud to do everything on site. This includes producing their own crops, with Scottish juniper, lemongrass, coriander and orris root on the roster of plants cultivated on site.

Always innovating, they've created a ground-breaking Rye Scotch Whisky, which is distilled from rye grown on their farms, reviving a style of whisky in Scotland that had disappeared. The team now also embrace gin tourism with their new Distillery Experience.

homage to the land as it stretches from the farm itself out to Lunan Bay on the east coast of Scotland, the gin uses kelp (the element of the sea), carline thistle and blaeberries (the rock and land).

Unlike the majority of UK-produced gin, Arbikie's premier gin uses a potato vodka made on the estate as its base instead of grain neutral spirit.

This gives it an 'extra smooth and distinctive' taste. This potato vodka base is macerated with juniper and the key botanicals before being put through AK and Jan, Arbikie's stills, named after the Stirlings' parents.

The Angus-based distillers then followed up the release of their original bottling with a second gin, dubbed AK's,

Named after the Gaelic word for nature, the world's first climate positive gin, Nàdar, is the brainchild of master distiller Kirsty Black. Distilled from the humble pea, with crops grown on the farm achieving a carbon-negative −1.54 kg of carbon per 70cl bottle, it is the culmination of a PhD in collaboration with the James Hutton Crop Institute and Abertay University on the effect of legumes in distilling.

Hoping to become the first climate positive distillery in the world, this gin is not only great for the environment but delicious to boot; silky smooth, the lemongrass and Makrut lime are complemented by the more classical gin botanicals.

53

GIN BOTHY

KIM CAMERON WILL be the first to admit that the Gin Bothy concept was created by accident when, following her original idea to enter a homemade jam into the World Jampionships in Perthshire in 2013, she discovered she had a surplus of fruit juice. A family member suggested she should use it to create a gin.

Kim's first attempt – made with an old traditional recipe – soon overtook the popularity of the jam and she quickly realised she was onto something with real potential. The jam Bothy soon became the Gin Bothy!

Kim decided to use a traditional method to create her gins. She follows Scotland's rich fruit calendar to make small batches of fruit liqueurs and a range of full-strength spirits.

The gin producer's first release was her Bothy Original gin, which is infused with botanicals including locally grown heather, milk thistle, hawthorn root and rosemary, as well as pine needles sourced from a local forest.

BOTANICALS
Juniper, cinnamon, cloves and spices

PREFERRED SERVE
Best served neat over some ice. Adding ginger ale will enhance the warmth of the cinnamon

PREFERRED GARNISH
A slice of orange

Gin Bothy, Kirriemuir and Kirkwynd, Glamis
ginbothy.co.uk
Price £36 / Quantity 70cl / ABV 37.5%

RECOMMENDED COCKTAIL
SMOKING GUN

Ingredients
50ml Gin Bothy Gunshot
1 dash of demerara syrup
3 dashes of Angostura bitters
2 dashes of orange bitters
Cinnamon stick and orange peel to
  finish

Method
1. Add the ingredients to a Boston shaker with ice and shake well.
2. Sieve through a strainer.
3. Serve with a cinnamon stick wrapped in orange peel.
4. Light the cinnamon stick for extra effect.

Their most popular creation is the Gin Bothy Gunshot Gin which is named after one of Scotland's oldest sports – shooting – and boasts a warming mixture of cinnamon, cloves and spices. Each bottle is hand-numbered, hand-batched and hand-poured.

Kim explained that it was originally created for a party of shooters and their hip flasks, who said they were after a tipple that would give them a much-needed 'inner glow' on the freezing cold moor and that also matched the colour of their bullets. Warming and filled with spiced notes, Gunshot is a savoury gin that's best enjoyed 'by the campfire'.

Keen to preserve bothy culture, Kim is known for singing bothy songs and is also a fan of sustainability, with a partnership with the Woodland Trust helping to offset the Bothy's carbon footprint and restore ancient woodlands.

Kim and the team have also launched the Bothy visitor site in the beautiful village of Glamis in Angus and expanded the core range to feature new gins and a top selection of fruit liqueurs.

MACKINTOSH

THE HUSBAND-AND-WIFE TEAM behind Mackintosh Gin say the idea that started their award-winning gin range began with a joke about how much gin they were buying each month. James joked to Deborah at the Wee G&T Festival down in Perth in 2016, as they perused more bottles, that it'd be cheaper to make their own.

One of the producers who overheard their conversation mentioned that was how they got started and the seed was planted. Eventually deciding to go ahead with it the pair took the first steps in Christmas 2017.

After researching and playing around with cold compounding to create some recipes, they approached contract distiller Lewis at Distillutions to tweak and perfect it before launching in October 2018.

Very much a family affair, James and Deborah's three daughters Stephanie, Charlotte and Alexandra, are all part of the brand, helping with development, social media and creating their innovative range.

Located in Angus, the family looked to the rich history of the County of Angus, where they make their home, and its connections to the Vikings, Picts and Celts, as well as its abundance of ancient standing stones to create the unique logo that marks their bottles.

Based on the 'Lover's Knot', originally carved into a Pictish standing stone in Meigle in the ninth century, the logo is made up of an unbroken line over four corners, the endless rotation of the seasons and the circle in the centre representing infinity, the sun, and the earth.

The family's first release was created using nine botanicals, the juniper complemented with specially selected botanicals including Mediterranean citrus fruits and locally foraged elderflower.

They followed their original up with a Mariner Strength, which elevates the original recipe to the higher ABV of 59% and is ideal for cocktails, and the Pineapple & Grapefruit Old Tom, which was designed with those with a sweeter tooth in mind.

Distillutions Distillery, Arbroath
mackintoshgin.com
Price £37 / Quantity 70cl / ABV 40%

Juniper, coriander seed, angelica, orris root, with dried orange, lemon and lime peels, as well as fresh pineapple, fresh grapefruit and elderflower. Candy syrup is added post distillation for sweetness

Over ice with lemonade

Pineapple, but grapefruit also works as does any other citrus fruit

MACKINTOSH'S WHITE LADY

Ingredients
50ml gin
25ml orange liqueur
25ml lemon juice
10ml sugar syrup
1 egg white
Lemon peel to garnish

Method
1. Pre-chill your glass in the freezer.
2. Pour all the ingredients into a shaker and shake well (dry shake).
3. Add ice to the shaker and shake well (wet shake).
4. Strain into chilled glass.
5. Garnish with lemon peel.

PINEAPPLE & GRAPEFRUIT OLD TOM

An unexpected success, the Pineapple & Grapefruit Old Tom was not only named Old Tom Gin of the Year at the Scottish Gin Awards 2020 but also beat out the other winners to take the top accolade of being Scottish Gin of the Year, too.

Speaking about how it came about, James stated that they had come across a lot of people at festivals who preferred other mixers such as soda water and lemonade over tonic and as such enjoyed the sweeter side of gins.

Deciding to create an Old Tom, an older, sweeter style of gin, the team experimented with different sugars, syrups and honey before settling on candy syrup which they felt worked best for what they wanted to achieve.

Adding fresh pineapple to their recipe which already contained fresh grapefruit, and combining it with the candy syrup, they created a deliciously sweet spirit that's quickly gone on to become their best-seller and for good reason.

REDCASTLE

BOTANICALS
13 in total including the usual suspects (juniper et al) and kaffir lime leaf, pink peppercorn, star anise, citrus peels, ginger, cardamom

PREFERRED SERVE
A simple tonic

PREFERRED GARNISH
Twist of lime

Toll House Spirits, Arbroath
redcastlegin.co.uk
Price £35 / Quantity 70cl / ABV 40%

WITH OVER 20 years' experience in the spirits industry, starting out as a rep for Gordon's (back in the early 2000s when gin wasn't as cool), Redcastle Gin founder Fiona Walsh decided it was time to make a gin herself and began developing their recipe when she was pregnant back in 2017.

A big fan of bolder-flavoured gins, she wanted to create her ideal spirit; smooth, clean, with fresh notes of citrus and bold spice. Reconnecting with contract distiller Lewis Scothern at Distillutions, the pair set about making Redcastle Gin a reality.

Released in July 2017, the London Dry Style Gin was an unapologetic, uncompromising, flavoursome gin that owed as much to Fiona's background in whisky as it does to the heightened senses of smell and taste she had due to pregnancy when they were developing the recipe.

A family affair, with her husband and sister also part of the team, the Redcastle Gin range, and the more accessible Toll

RECOMMENDED COCKTAIL

REDCASTLE SNAPPER

Ingredients
Spice sachet by Little Devil
50ml Redcastle Gin
120ml tomato juice
A wedge of lemon

Method
1. Add the spice mix, gin and tomato juice to a glass over ice.
2. Stir and garnish with the lemon wedge.

House Gin range, which also make up part of their portfolio, are named in nod to her late mother, as both the Lunan Bay landmark and the Toll House in Brechin are connected to her.

The team also produce the Redcastle Passion Fruit & Mango Gin, Redcastle Blood Orange & Rhubarb Gin, and Redcastle Raspberry & Pomegranate Gin, as well as a range of tasty gin liqueurs.

REDCASTLE GIN

As ideal for drinking neat as it is for experimenting with, this London Dry Style Gin is filled with layers of flavour due to its punchy recipe of 13 botanicals in total including the likes of kaffir lime leaf, pink peppercorn, star anise, citrus peels, ginger and cardamom.

Combining to create a rich gin that delivers on the nose as well as the palate, each bottle is individually filled, sealed, wax dipped, labelled and numbered by hand in Arbroath, Angus, with Fiona succeeding in her aim to create a spirit that was filled with big hits of citrus and spice.

Expect this one to divide opinion, but with the followers of the Scottish Gin Society regularly ranking it in their top ten for the annual Scottish gin poll, the polished, clean taste, punchy flavours and mellow citrus finish clearly score high with Scotland's gin fans, and I'm sure you'll agree.

CENTRAL SCOTLAND

THE GINS

NORTHERN SCOTLAND
HIGHLANDS, SPEYSIDE & ABERDEENSHIRE

56

ALEXANDER'S

BOTANICALS
11 botanicals including juniper, orange and Royal Deeside honey

PREFERRED SERVE
A light tonic and ice

PREFERRED GARNISH
A slice of pink grapefruit

Bigfish Brewing Co, Stonehaven
alexandersgin.com
Price £38 / Quantity 70cl / ABV 44.4%

THE STORY OF Alexander's Gin begins in the Caribbean where founder Fred Stockton was offered an orange-based spirit by one of the local distillers.

Inspired by the taste and the belief he could improve on this flavour as a gin, the homebrewing enthusiast, who has been brewing under the Big Fish Brewing Co name since 2002, became fascinated with the idea of creating his own spirit.

After a chance meeting with the guys at Lost Loch Spirits, he began the first steps that would lead to the creation of Alexander's.

Working with the Lost Loch team, he began cuckoo-distilling his new gin with the experienced distillers on hand right from the original concept to the recipe development and to Fred bottling the batches himself by hand.

Though it took a lot of tinkering with the recipe, the Stonehaven-based drinks enthusiast said he was delighted by how accommodating LLS were and that they've ended up with a gin they are all really proud of, with the traditional London Dry juniper-led spirit building to a wonderful citrus orange-filled finish.

When it came to the runic design aspects for the bottle, Fred said he was influenced by his Orcadian roots. He combined the Old Norse symbol for 'where there's a will there's a way', the double X on top of one another, with the letter 'A' for Alexander (his daughters' suggestion) to create the final logo.

The name itself goes back to Fred's experience in the Caribbean. Fred jokes that to this day he has never found out the name of that Caribbean distiller, so in honour of the fact he helped start this incredible journey, Fred named him 'Alexander' and dedicated this wonderful new gin to him.

ZESTY MEARNS

Ingredients
50ml Alexander's Gin
50ml sugar syrup
25ml lime juice
1 tsp orange thin cut marmalade
1 tsp very finely cut fresh ginger

Method
1. Combine all ingredients and dry shake until marmalade dissolves.
2. Add crushed ice and shake vigorously for 10 seconds.
3. Loosely strain into a Martini glass and then top up with the original crushed ice and 'good bits' left from the shaker.

Dunnottar Castle, Stonehaven

NORTHERN SCOTLAND

57
CITY OF ABERDEEN

BOTANICALS
Bilberries, pink peppercorns, lavender

PREFERRED SERVE
Over ice and mixed with light tonic water (1:2)

PREFERRED GARNISH
Sliced strawberries or a slice of apple or sliced grapes

City of Aberdeen Distillery & Gin School, Aberdeen, Scotland
cityofaberdeendistillery.co.uk
Price £34.99 / Quantity 70cl / ABV 42%

ONE OF THE regions in the country that really deserves more focus on its burgeoning food and drink scene, Aberdeenshire has seen a rise in excellent spirits and beers that's making the rest of the country sit up and take notice.

Strangely, Aberdeen itself has not had a working still in living memory, but, thanks to the work of two spirits-loving friends, Alan Milne and Dan Barnett, that has now changed. The duo had their 'wouldn't it be a great idea if . . .' moment after meeting at Aberdeen University's student wine appreciation society and the Granite City now has its first legal distillery for nearly 80 years.

Alan, who hails from Buckie and comes from a wine-making background, and Dan, from Manchester, with a deep love for home-brewing, then founded the City of Aberdeen Distillery and Gin School.

After a few setbacks, the pair set up shop in a former storehouse in an arch under Aberdeen's main railway line, where they offer a trio of tours, tasting masterclasses and gin-making experiences, as well as providing an ever-changing line-up of bottle your own gins that can only be bought from the distillery.

All of their spirits are made using premium organic alcohol, including their Fresh Gin (LDG) which is freshly made every two to three weeks using carefully prepared citrus, namely grapefruit and orange.

ABERDEEN GIN

Not wanting to force a gin flavour onto Aberdeen, the pair instead decided to ask the city to choose the flavour itself. Launching tasting packs at the Taste of Grampian, gin fans were able to vote on the final recipe for Aberdeen's Gin.

The result is a fruity, lightly floral with refreshing citrus gin that comes in a bottle with a stunning label, created by local design agency FortyTwo Studio, that tells the story of Aberdeen, featuring illustrated references from the city's fishing and rope-making heritage to more modern influences like the mortar-board (for the city's top universities) and, of course, the seagulls.

SCOTTISH STRAWBERRY SMASH

Ingredients
50ml Aberdeen Gin
3 fresh Scottish Strawberries (keep one aside for a garnish)
1 lime wedge
½ tsp of granulated sugar
Soda water

Method
1. Using a tall glass, add the sugar and a squeeze of lime.
2. Dissolve the sugar by muddling / stirring.
3. Slice and add two of the strawberries and lightly muddle.
4. Fill the glass with ice, add the double measure of Aberdeen Gin, top up with soda water.
5. Garnish with a strawberry and a sprig of mint.
6. Enjoy!

LONEWOLF

BOTANICALS
Juniper, Scots pine, fresh grapefruit peel, fresh lemon peel, coriander seed, cardamom, angelica root, orris root, Thai lemon grass, pink peppercorn, kaffir lime leaf, mace, almond and lavender flower

PREFERRED SERVE
Over ice in a 1:3 ratio of LoneWolf Gin to a premium tonic

PREFERRED GARNISH
A twist of pink grapefruit peel, pith removed

BrewDog PLC, Ellon
brewdog.com/uk/lonewolf-gin
Price £25/ Quantity 70cl / ABV 40%

THESE DAYS NORTHEAST Scotland has a lot to shout about when it comes to food and drink, not just that which is found on its farms and fishing boats, but also in terms of what its breweries and distilleries produce.

Home to one of the country's – and arguably the world's – biggest craft beer producers as well as some world-class single malt whisky distilleries, BrewDog can now add gin to its long list of potable delights.

The spirits arm of craft beer giants BrewDog was launched in April 2017 and is one of only a handful of grain-to-glass distillers in the country.

The team at Ellon controls every aspect of their gin's production. The brand's own vodka is made with a 50:50 blend of malted wheat and malted barley.

It's created in the specially designed distilling equipment – which features a 19-metre high column – before the botanicals are added to create their signature London Dry Gin.

The one local botanical, of the 14

they use, is the needles of the Scots pine, a tree that grows in abundance in Aberdeenshire and imparts a unique flavour to their London Dry gin.

Alongside the London Dry, BrewDog's spirits arm also produces a Navy Strength Gunpowder Gin, which they launched in 2017, a vodka, several whiskies, a rum and a range of full-strength flavoured gins including the refreshing Cloudy Lemon, unique Cactus and Lime, and newly launched Peach and Passionfruit.

Never ones to hold back when it comes to the pursuit of bright, shiny things, the LoneWolf team have even taken the hassle out of choosing a tonic by releasing a canned G&T that has the perfect ratio of LoneWolf London Dry and their own specially created tonic water.

RECOMMENDED COCKTAIL

LONEWOLF NEGRONI

Ingredients
60ml LoneWolf Gin
30ml Campari
30ml sweet vermouth
Valencia orange peel, to garnish

Method
1. Put all the ingredients in a mixer with ice and stir until well chilled.
2. Strain into a rocks glass, add a large cube of ice and garnish with a twist of Valencia orange peel.

LONEWOLF LONDON DRY

Described as 'purity personified' and a 'gin with bite', LoneWolf London Dry was created after a long series of experiments to formulate a perfectly balanced recipe designed to maximise flavour.

Featuring 14 botanicals, LoneWolf begins with a body of citrus giving way to a delicate spiciness before the lavender provides a wonderfully floral finish. This is a gin with depth and complexity that rewards the drinker with its different qualities at every sip.

EENOO

BOTANICALS
Royal Deeside honey, heather flowers, brambleberries, raspberries, Italian juniper, rose hip, coriander seeds, angelica root, liquorice root, almonds, pepper, orange and lemon peel

PREFERRED SERVE
In a glass with ice, topped up with lite tonic

PREFERRED GARNISH
Diced strawberries

The Lost Loch Distillery, Aboyne
lostlochdistillery.com
Price £35 / Quantity 70cl / ABV 43%

HEAD WEST FROM Aberdeen and follow the winding River Dee through the rolling hills and sprawling rural scenery of Royal Deeside, and you'll eventually come to one of the most whimsical spirits producers Scotland has to offer.

Lost Loch Spirits was set up by Peter Dignan and Richard Pierce in 2017 and is named after Loch Auchlossan.

The spirits producer can be found on the historic eastern shore of this 'lost' body of water. Lovers of alcoholic drinks and their history, Dignan and Pierce's pioneering claim to fame is that they are the first distillery team in the country to have produced their own absinthe, Murmichan, which takes its wonderfully evocative title from the name of a wicked Scottish fairy.

Their small-batch gin, which followed shortly after Murmichan, is called eeNoo after the old Scots phrase meaning 'just now', 'at the present time', or 'at once'.

Designed to be a traditional gin modelled on the juniper spirits of old,

LOST LOCH

Ingredients
50ml eeNoo
10ml Haroosh
2 dashes of Murmichan
30ml fresh orange juice

Method
1. Add all the ingredients to an ice-filled shaker and shake for a good 20 to 30 seconds.
2. Strain into a tall glass.

eeNoo also features a selection of botanicals sourced locally.

The spirits range, the majority of which is produced on a 500-litre still and a smaller traditional copper alembic still, now includes new addition Haroosh – a brambleberry and whisky liqueur.

All of the production, bottling and labelling still takes place at their Highland distillery.

In addition, the pair have gone to great lengths to use sustainable energy sources; a percentage of the power and heat used at the distillery comes from renewables such as a wind turbine, solar panel array and biomass boiler – all located on-site.

They've also just launched the Lost Loch Spirits School which gives aspiring distillers access to their individual copper stills and intriguing Wall of Botanics to create their own signature gin or even an absinthe.

EENOO

The eeNoo recipe is filled with botanicals designed to evoke the Scottish Highlands at each taste. The most predominant of these is Royal Deeside honey, which has a flavour profile unique to the region with heather, willow herb and clover pollen all contributing to its distinctive make-up.

Heather flowers and soft fruits such as brambleberries and raspberries sourced from berry farms in Aberdeenshire, Deeside and Angus are the perfect complimentary flavours.

This leads to a traditional juniper flavour coupled with notes of Scottish berries and a lingering spice. The result is then lovingly bottled by hand and given a striking label featuring an enigmatic Inuit traveller named Eenoolooapik who, according to the distillers, visited Scotland, and in particular the city of Aberdeen, in the early nineteenth century.

NORTHERN SCOTLAND

ESKER

BOTANICALS
Juniper, pink peppercorn, citrus, cassia, heather, rose hip, milk thistle and silver birch sap

PREFERRED SERVE
Over ice with a good tonic. However, the Esker team encourage people to experiment. The gin can also be taken neat over ice, or mixed with cloudy lemonade or a ginger ale, then garnished with pink grapefruit, cinnamon stick or rosemary

PREFERRED GARNISH
A twist of orange peel

Esker Spirits Ltd, Aboyne
eskerspirits.com
Price £35 / Quantity 70cl / ABV 42%

THE STUNNING SETTING of Royal Deeside, and the valley of the River Dee, is home to the summer residence of the royal family, Balmoral Castle, but also to a brilliant new gin producer.

Esker Gin is the brainchild of Steven and Lynne Duthie, a couple who grew up in the North East and wanted to create a gin brand that would reflect the stunning vistas of their local surroundings.

Having founded the distillery in October 2015, the pair went on to launch their first gin in June 2016, featuring an intriguing, locally sourced botanical unique in Scotland to Esker: silver birch sap.

Before that, the distillery began life as a series of experiments in a small still in their kitchen, which then emigrated to a shed in the garden before moving to the Kincardine Estate in 2017 – an upscale that was due both to increased demand for the gin and to gain easier access to their unique botanical.

Chosen for its curiously sweet flavour, the silver birch sap is sourced

APPLE CHARLOTTE

Ingredients
30ml Esker Gin
15ml vermouth
30ml pressed apple juice
30ml ginger beer
Crystallised (candied) ginger, to
  garnish

Method
1. Mix all the ingredients in a tall glass over ice
2. Garnish with the crystallised (candied) ginger.

their empty bottle at events in return for a discount on their purchase.

Their zero-waste policy means that effluent (waste product from stills) is now given to the estate's farmers as fertiliser.

The Duthies have expanded their drinks range beyond the two gins they originally launched with – the Esker and the Esker Gold Gin – by adding the Scottish Raspberry, Esker Valencian Orange and Esker Silverglas London Dry Gins, as well as a range of naturally flavoured fruit vodkas.

from the trees of the Kincardine Estate, which along with heather flowers and rose hip, provides the provenance the pair were keen to create when establishing the brand.

Determined to build a company they could be proud of, the Duthies have put people and sustainability at the heart of their operation, with anyone who works for Esker, be it full-time, part-time, temporary or ad-hoc event staff, paid significantly over the national living wage, regardless of age.

In addition, they were one of the first in the craft spirits industry to introduce a scheme where customers can return

ESKER GIN

Designed to be a classic juniper-led gin but with hidden depth, Esker takes a little sweetness from its silver birch sap as well as floral notes from rose hip and heather.

The recipe took over two years of experimentation to perfect with the Scottish botanicals added to traditional ingredients sourced elsewhere, such as pink peppercorns and cassia, to create a complex gin full of appeal to traditionalists.

The modern design of the bottle features references to the River Dee, the Deeside topography, the silver birch and, of course, a hint of tartan.

NORTHERN SCOTLAND

61

HRAFN GIN

BOTANICALS
Juniper, angelica, cassia, coriander,
cubeb pepper, orris root and mandarin

PREFERRED SERVE
Over ice with a light tonic

PREFERRED GARNISH
A slice of lemon

Burn o'Bennie Distillery, Banchory
hrafngin.com
Price £39 / Quantity 70cl / ABV 45%

A BERDEENSHIRE'S WELLSPRING
of great gins shows no signs of
running dry and the inventiveness and
entrepreneurial spirit rife in the region
has given rise to yet another excellent gin
in the form of HRAFN GIN.

The story of HRAFN is a saga that
involves two whisky-loving brothers,
Peter and Callum Sim, and their journey
to create premium Scottish gins that are
deeper in taste, longer in finish and so
smooth you can enjoy them neat.

The result was Thought & Memory,
named for Huginn and Muninn who were
the Norse God Odin's two ravens.

This delightful well-balanced gin is
made using seven carefully selected
botanicals including the subtly sweet
mandarin.

The brothers followed up this exciting
first release with 'Valhalla', a stronger,
more intense version of 'Thought &
Memory' and 'Winter', a seasonal release
that features the botanicals frankincense
and myrrh.

Combining one of Scotland's most

famous Scottish desserts with their expertly made spirit, they released Cranachan early in 2021. HRAFN GIN's first foray into flavoured gin, it's made using locally sourced raspberries, honey, oats and new-make malt spirit.

Also new for 2021, is Valkyrie, a London Dry gin that uses jara: a semi-wild citrus fruit. This extended range now means HRAFN can cater for all tastes with their innovative range.

HRAFN GIN THOUGHT & MEMORY

Fresh on the nose, light and smooth on the palate, with a distinctive warm and slightly spicy finish, the unmistakable sweet notes of the mandarin are carried throughout Thought & Memory.

The stylish bottle, replete with the eponymous ravens, is offset by copper styling on the lettering, and a customised copper stopper, to reflect the slow distillation involved in the gin's creation.

RECOMMENDED COCKTAIL
MANDARIN MOJITJA

Ingredients
50ml HRAFN Gin Thought and Memory
100ml Fever-Tree Sicilian lemonade
7 mint leaves
½ unpeeled mandarin, chopped
10ml sugar syrup

Method
1. Add the mint and mandarin to a heavy-based rock glass and pour over the sugar syrup.
2. Muddle the mint and mandarin and fill the glass with ice.
3. Pour over the gin and top with the lemonade.
4. Garnish with mint leaves and mandarin.

Little Brown Dog Spirits, Aberdeenshire
littlebrowndogspirits.com/lbdgin
Price £32 / Quantity 50cl / ABV 43%

LBD GIN

WHILE MOST GIN producers get dogs to go with their distilleries, Andrew Smith, co-founder of Little Brown Dog Gin, jokes that they got a distillery to go with their dog.

He and fellow founder Chris Reid, who were originally brought together due to a love of rallying and mountain biking, decided to use an outbuilding on Chris's farm to launch a new spirits distillery – complete with a 100-litre copper pot – in 2018.

The exciting new firm was based on Andrew's experimental LBD brand, which is named after his pet dog Banksy, and under which he'd created small-batch releases since 2013 for him and his friends.

Andrew jokes that he is the ideas man while Chris is the practical one who comes up with the solutions as to how to make things actually work on their tight budget.

Their first major release was a gin which they named Project AFG (short for Another F**king Gin, although they told

Chris's mum it meant Aberdeenshire Foraged Gin), designed with an aim of creating a spirit that was a fingerprint of the northeast region in which it is made.

The first-ever 'unlimited edition' that moved Andrew's experimental spirits to a more permanent range, the pair wanted to create a recipe that used sustainably foraged botanicals and would be capable of being upscaled if needed.

The LBD Gin was the next step in this evolution, a combination of their Project AFG and GPS it featured a streamlined recipe that played up to the oily spirit they create with their small still, yet it also stayed true to their 'forage what we can, grow what we can't and only buy what we cannot grow or forage sustainably' ethos.

This transparency shines through on their website where, not only can you find out which botanicals they use, but also where they come from as well as a what3words location for the foraged and grown in-house ingredients.

As well as the LBD Gin core gin, the

Birch sap, juvenile beech leaves, wood sorrel, parsnips, rhubarb lemons, grapefruit, coriander, orris root, angelica and bee pollen

Created for experimentation, Andrew and Chris say you should drink it any way you like

Having used intriguing garnishes like local Aberdeenshire Asparagus, foraged sorrel and wild lilac in the past, the duo recommend being as inventive as you like

RECOMMENDED COCKTAIL
LBD NEGRONI

Ingredients
35ml LBD Gin
35ml sweet vermouth
25ml Sweetdram's whisky amaro
Twist of orange

Method
1. Stir all three ingredients over ice. Andrew adds that anyone looking to go 'full hipster' with it should then infuse with birch bark smoke, preferably from one of the geotagged birch trees that they tap for sap.
2. Finally, serve with a twist of orange.

pair have launched a Latitude Strength version of their LBD Gin which comes in at a whopping 57.2% ABV; distilled at 57.2° north and bottled at latitude strength.

LBD GIN
A modern dry Scottish gin, LBD Gin is created using Scottish botanicals that are foraged from within the vicinity of the farm, including birch sap, which Andrew and Chris tap themselves during a two to three week window in spring.

The sap is then frozen to use for the rest of the year. Sweet juvenile beech leaves emerge in May and the duo also freeze these for future use.

The recipe features a fair whack of juniper, with 5% of the bill coming from Scottish stock. Since 2019, they have been planting juniper around the farm in a bid to become fully self-sustainable in the essential ingredient.

The pair grow the parsnips and rhubarb, while the bee pollen is now produced by LBD's own hives which are set up within walking distance of the still.

VESPERIS

DEEP IN THE heart of what was once known as Pictland – that's the North East to you and me – lies a craft distillery that has gone back all the way to Roman times in search of inspiration for their gin.

Blackford Craft Distillery, a family-run production site, set up by Neil and Katie Sime in 2017, is based in a nineteenth-century steading that was once part of the Blackford Estate, with links to Fyvie Castle.

Maryfield Steading had been neglected until the Sime family took it over and sympathetically refurbished it into a still room and fully licensed distillery.

After launching their first spirit – Pictish Vodka – to great success, the Simes turned their attention to gin. They created their first offering using an ancient Pictish recipe for heather honey mead, which was originally brewed on the slopes on Bennachie, Aberdeenshire.

Designed to be a modern gin infused with 'ancient inspiration', Vesperis Pictish Gin uses the heather honey and heather

BOTANICALS
Apples, heather honey, heather blossom and juniper berries

PREFERRED SERVE
Over ice, served with premium tonic water

PREFERRED GARNISH
Red fruit, particularly pomegranate or frozen raspberries

Blackford Craft Distillery Ltd, Inverurie
blackfordcraftdistillery.co.uk
Price £35 / Quantity 70cl / ABV 40%

SPRINGTIME SOUTHSIDE

(courtesy of The Wandering Bartender)

Rhubarb and mint really complement Pictish Gin without overpowering it. The subtle flavours of the juniper berries and coriander really come through along with the heather blossom finish, which is what gives this gin its special edge.

Ingredients
50ml Vesperis Pictish Gin
25ml lime juice
12.5ml sugar syrup
5 mint leaves
4 cubes of rhubarb

Method
1. Muddle the rhubarb and sugar syrup in a cocktail shaker.
2. Measure gin and lime juice into a cocktail shaker, add ice and shake.
3. Double strain into a rocks glass with ice and garnish with a sprig of mint and rhubarb peel.

blossom prevalent in the mead recipe as its primary botanicals. Their newly planted orchard of a hundred heritage apple trees provides the apples for their gin and vodka.

The team have big aspirations. In 2018, they launched Scotland's first honey vodka – Vesperis Heather Honey Vodka, which uses Aberdeenshire heather honey from Udny Provender in Methlick and is based on a traditional Polish honey vodka, krupnik.

Vesperis Pictish Gin is made using local botanicals including raw heather honey, which is sourced from a local beekeeper, Udny Provender.

This is then combined with hand-picked heather blossom from Royal Deeside and Bennachie, as well as seasonal apples from their orchard in Aberdeenshire.

Together these create a 'completely unique' gin in which citrus and spice compete with a full floral finish from the heather.

The Picts historically observed the planets to track the seasons and they would carve their astronomical calendar – the V-Rod and Crescent – onto their standing stones.

To represent this, the logo on the bottle was designed with Mither Tap – one of the highest peaks of Bennachie – in the centre with Venus as the 'evening star' above it.

The gin itself takes its name from 'Vesper', the ancient name for the evening star.

INSHRIACH

SET DEEP IN the heart of the Cairngorms National Park and located between the River Spey, Loch Insh and the foothills of the Cairngorms, in 200 acres of woodland, Inshriach House lends its grounds and its name to an enticing Speyside gin.

The distillery itself came to the attention of the public in 2015 when it won Channel 4's Shed of the Year and was initially the home of Crossbill Gin. In 2017 Inshriach relaunched with its own Original and Navy Strength gins.

Growing organically, they've since added a bottling room and bonded warehouse to bring the entire production process in-house, as well as moving into product development and contract distilling, helping to produce both Fidra and Duncan's gins.

A new building is in the pipeline to allow the business to expand further and a giant biomass system is being constructed to power not only the distillery but the whole estate.

Originally founded by Walter

Micklethwait, Inshriach is a one-man operation. Walter enlists the help of volunteers when it comes to harvesting time because the window when the ingredients are 'just right' is relatively narrow.

The region is one of the few remaining strongholds of Scottish juniper which grows all around Inshriach, with Walter freezing the berries to keep their flavour.

Theoretically, all the ingredients to

NORTHERN SCOTLAND

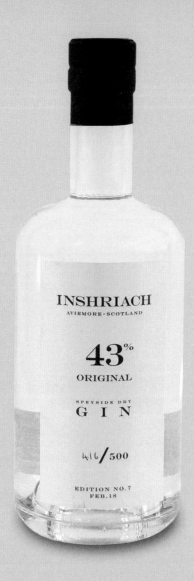

Inshriach Distillery, Aviemore
inshriachgin.com
Price £35 / Quantity 70cl / ABV 43%

INSHRIACH'S WHITE LADY

Ingredients
35ml Inshriach Gin
10ml Cointreau
25ml lemon juice
10ml sugar syrup
Lemon peel, to garnish

Method
1. Fill the shaker with ice and add the ingredients.
2. Shake until cold then strain into a coupe glass.
3. Garnish with a twist of lemon peel.

make the gin could be picked within a few hundred yards of the distillery but Walter has found favourite patches of juniper up and down the Spey Valley.

Only two other botanicals are then collected: the rose hip, which is found close to the juniper in the Spey valley, while the Douglas fir is taken from the nearby forest.

Even the water is sourced from a spring on the hill behind the estate adding its own distinctively fresh flavour. The shed in which the stills are located is reminiscent of a frontier building from a spaghetti western and

is locally sourced and mostly recycled, making Inshriach an extremely low-impact business.

The distillery is not normally open to visitors, but Walter says community is important to him so he holds regular open days – or 'mini-festivals' – where people can enjoy food, music, gin and plenty of cocktails.

They also host botany and foraging weekends once or twice a year, allowing their fans to explore the grounds and learn all about the pivotal plants and berries that go into Inshriach's making.

INSHRIACH ORIGINAL GIN

Made in very small batches of just 500, the gin's label is beautifully simple, reflecting the rustic nature of the distillery itself. The ABV is writ large on the front and the batch number hand-written on the bottom.

With evocative floral sweetness and strong hints of that fresh Scottish juniper, Walter describes his gin as 'the equivalent of walking in the forest – fresh and clear and airy but juniper-forward with pine and sweetness'.

The three components offer their own flavours in an excellent balance, and this makes it perfect for drinking neat over ice though it is robust enough to make an excellent gin and tonic.

KINRARA GIN

BOTANICALS
Juniper, coriander, lemon peel, sweet orange peel, liquorice root, angelica root, rose hips and rowan berries

PREFERRED SERVE
Large measure of Kinrara Highland Dry with twice as much Fever-Tree Mediterranean tonic and lots of ice

PREFERRED GARNISH
A twist of orange peel

Kinrara Distillery Ltd, Aviemore
kinraradistillery.com
Price £38 / Quantity 70cl / ABV 41.5%

LYING CLOSE TO the River Spey and the mountaineers' paradise that is Aviemore, the scenic backdrop of the Cairngorms National Park and soaring peaks of Ben MacDui provide the inspiration for this exciting gin distillery.

Sporting one of the best views from their front door, Kinrara was built to 'breathe life back into a 200-year-old milking steading' on the Kinrara estate.

The surrounding area offers an expansive range of flora and berries that Kinrara hand forage for their gins, mixing them with carefully sourced botanicals from around the globe. The distillery is keeping a hands-on approach to the distilling too, with all of the cut points decided by their head distiller David Wilson using nosing and tasting of the spirit.

Their flagship gin, Kinrara Highland Dry Gin, has gone down a storm since its launch, winning multiple awards including being named 'Best Contemporary Gin' in Scotland at the 2021 World Gin Awards.

The brand is also keen to support local charities, with a £5 donation from each bottle of their Cairngorm Whiteout Gin going to support the Cairngorm Mountain Rescue Team. There is currently a distillery shop on site, with further plans to expand the distillery to include a visitor centre and tasting room to ensure that gin lovers can fully enjoy the Kinrara experience.

KINRARA HIGHLAND DRY GIN

Produced on two handmade solid copper pot stills from Portugal, with all of the cuts done by taste not yield, Kinrara Highland Dry Gin is described as a juniper-led dry gin, combining foraged and locally sourced botanicals such as rose hips, rowan berries and sweet orange peel.

Floral and sweet, this a great gin, well worth trying either on its own with ice and an orange twist garnish, or mixed with a tonic. Each batch of 240 bottles are individually numbered and come with stunning packaging that reflects the glorious scenery that surrounds the distillery, with a striking image depicting the Cairngorms as seen from Loch Morlich.

RECOMMENDED COCKTAIL
SCOTTISH SUMMER

Ingredients
30ml Kinrara Highland Dry Gin
10ml tropical oleo
10ml lime juice
15ml Aperol
Soda

Method

For the oleo
1. Peel your citrus fruit with a knife or peeler, ensuring as much of the white pith is removed as possible.
2. Add the peels to a mason jar and cover with caster sugar, using roughly 50g per lemon as a guide (scale up if using larger fruit).
3. Seal the jar, shake and leave overnight. The next day, the oil from the peels will have formed a syrup.
4. Add 50ml boiling water per citrus fruit to dissolve the remaining sugar and lengthen the syrup.
5. When all the sugar has been dissolved, remove the peels, bottle and refrigerate for up to a week.

For the cocktail
1. Combine all the ingredients (other than the soda) in a shaker and shake well.
2. Strain into a tall glass with ice and top with soda.

Moray Distillery Ltd, Elgin
avvascottishgin.co.uk
Price £35 / Quantity 70cl / ABV 43%

AVVA

THE SPEYSIDE REGION is, of course, world-famous for its whisky production. However, thanks to the Elgin-based Moray Distillery and Avva Gin – launched by drinks industry veteran Jill Brown in 2016 – the area now holds its own in Scotland's burgeoning gin scene.

The creation of the distillery came at the end of a six-year journey, which, Brown explains, was the result of a long time spent planning and building up the courage to action those plans. A one-woman show, the Moray Distillery is, much like the fifty or so whisky distilleries that neighbour it, very much rooted to the land and community in which it is based.

Moray is one of the only Scottish gin distilleries to have a Scottish-made still – which was specially crafted at the Speyside Copperworks. Brown named this bespoke copper still Jessie Jean – or J.J. for short – after her grandmothers.

The Speyside connection can also be seen in the brand's choice of botanicals. Nearly half of those used to make Avva Gin are foraged in the area and, perhaps

BOTANICALS
Juniper, coriander seed, angelica root, orris root, rowan berries, dandelion, red clover, mint, nettles and citrus peel including lemon and orange

PREFERRED SERVE
Straight over ice

PREFERRED GARNISH
A slice of red apple

BASIC GARNISH
A slice of lime

more poignantly, all were 'weeds' the distiller says she originally found growing on her grandparents' farm.

And, of course, no Speyside venture would be complete without a connection to the spirit that is so deeply rooted in the region. Thus, in 2017, Moray Distillery released a gin matured in a whisky cask.

AVVA BAX-STAR MARTINI

This is a twist on the Pornstar Martini, which uses the Mango & Passion Fruit Curd made by Avva Gin's neighbours, Baxters.

Ingredients
35ml AVVA Gin
1 good heaped tsp of Baxters Mango
 & Passion Fruit Curd
Squeeze of lime
Tonic or prosecco

Method
1. Load a shaker with ice and the gin, curd and lime – shake hard and fast to ensure the curd mixes through.
2. Strain into a chilled martini glass and top with tonic or prosecco.

The former bourbon cask used to age the special gin was chosen in clear homage to Speyside whisky; many of the distillery's neighbours use these particular barrels to mature their new-make spirit.

The name Avva was chosen for the fact that it means 'respected grandmother' or 'respectable elder woman' in the Indian language of Dravidian, while in Hebrew it apparently means to 'overturn or ruin', which is a cheeky nod to the spirit's age-old moniker 'mother's ruin'.

The distillery now offers a tasting tour, where fans can not only see how this exciting gin is made but also get to try a sample flight of their exciting gins which include the Navy Strength, Pink and recently-released colour-changing Blink Gins.

Avva Gin is known for being very smooth and, like its neighbouring single malts, can be drunk over ice without a mixer. A London Dry style gin, it's a 'clean finish, well-balanced' example of the style, which has proved hugely popular since its launch in 2016.

A single shot, small-batch – 120 litres – and vapour-basket-distilled gin, Avva is created using 11 botanicals with five of them coming from the area around the distillery.

The dandelion, nettle, mint, red clover and rowan berries – which Brown says comes from her friend's farm outside Dufftown – are all foraged from Caithness down to Speyside.

The focus and careful planning that went into creating Avva's flagship spirit can also be seen in the branding. The bottle that carries the gin was created in tribute to a local landmark once known as the 'Lantern of the North', the Elgin Cathedral. The logo design references the building's famous rose window – which also appears on the large cork stopper – and the arched windows form the label's background.

CAORUNN
{KA-ROON}

PROOF OF THE idea that you don't have to be a new distillery to take advantage of being able to make both whisky and gin, Speyside's Balmenach Distillery brought specialist equipment to their site in 2008 before beginning production of the juniper spirit alongside their whisky in May 2009.

Forming part of the big three Scottish craft gins alongside the Botanist and Hendrick's, Caorunn – pronounced 'ka-roon' – played a huge part in driving this category forward and it should take a lot of the credit for its success. 'Caorunn' is the Gaelic word for rowan berry, one of the gin's most important botanicals.

Gin Master Simon Buley, who also happens to be the assistant manager at Balmenach Distillery, explained that the link with whisky production helps in some ways – such as nosing, tasting and quality control – but the distilling technique used to produce the gin is so unique that it takes a fair amount of adjustment to master it. The specialist equipment – supplied by the former owners of Inver House Distillers – comes in the form of an original 100-year-old copper berry chamber.

This has been adapted from its original use – which was to extract essential oils as a base for perfumes. Buley believes that their unique system of vapour infusion, which sees the 11 botanicals placed in four baskets within the copper still, helps to create a light, supremely balanced spirit with the copper contact helping to strip away some of the harsher flavours. Buley's aim is that Caorunn should 'distil the Scottish Highlands into a glass': it's a drink made to directly reflect the area in which it is produced.

The distillery offers a behind-the-scenes tour of their production area, along with a tutored deconstructed nosing and tasting session, finishing off with a refreshing Caorunn and tonic with crisp red apple wedges in their innovative bothy.

They've also recently added their Highland Strength and Scottish Raspberry Gins to their core range.

Balmenach Distillery, Grantown-on-Spey
caorunngin.com
Price £29 / Quantity 70cl / ABV 41.8%

BOTANICALS

Juniper, coriander seeds, lemon peel, orange peel, angelica root, cassia bark, heather, rowan berries, Coul Blush apples, bog myrtle and dandelion leaves

PREFERRED SERVE

Gin and tonic

PREFERRED GARNISH

A slice of red apple

RECOMMENDED COCKTAIL
APPLE WINTER TODDY

Ingredients
45ml Caorunn Gin
125ml cloudy apple juice
Dash of port
20ml lemon juice
2 tsps caster sugar
2 dashes of Angostura bitters
Orange slice with cloves, to garnish
Slices of red apple, to garnish
Grated nutmeg, to garnish

Method
1. Add the ingredients together, stir with a spoon then heat in a pan over a stove for a few minutes.
2. Once hot but not boiling, pour into a glass and garnish with a clove-studded orange slice, a few slices of red apple and grated nutmeg.

CAORUNN GIN

Caorunn is made using 11 botanicals; five of which are specifically selected owing to their use throughout Celtic history for medicinal purposes. All five of these locally sourced wild botanicals grow within a five-minute walk of the distillery.

The team use heather, which grows on the hills around the distillery; Coul Blush apples, which originated in the 1800s in Coul, Ross-shire; bog myrtle, which grows next to the heather and, rather delightfully, is also used for midge repellent; dandelion leaves and rowan berries.

These are added to the juniper and five other traditional botanicals to create a clean, crisp, fruity and slightly floral gin that is well rounded with no one flavour taking too much of a lead.

NORTHERN SCOTLAND

Glenrinnes Distillery, Keith
eight-lands.com
Price £39 / Quantity 70cl / ABV 46%

EIGHT LANDS

NESTLED AT THE foot of Ben Rinnes, right in the heart of what is Scotland's most industrious whisky-making region, you'll find one of the country's most exciting new gin producers.

Named for the eight counties that surround the iconic mountain in Speyside, Eight Lands Gin was developed by father and stepson team, Alasdair Locke and Alex Christou, who wanted to highlight the incredible water that flowed from the springs on their land and the wealth of botanicals in the countryside that surrounded them.

Originally forming the idea in 2016 after Alex suggested they bottle their own spring water, a chance discussion about the botanicals that could be found in the Speyside region steered the family towards the idea of making a spirit that would combine both.

Commissioning a botanical survey to find which plants they could take from their estate without damaging the local ecology, using the fact that both Alex and Alasdair preferred a classic London Dry style gin, the pair set about creating 15 different recipes that they then honed down using feedback from a selection of top London bartenders.

Alex confirmed that the final tweak was to change the AB from 44% to 46%, which their bartender collective felt would see the flavours cut through in cocktails better, and saw their 16th iteration go on to become Eight Lands Organic Speyside Gin.

With Glenrinnes Distillery being built in 2018, the pair alongside Meeghan Murdoch, the head of distilling and operations, began distilling in the early part of 2019 before launching both their Eight Lands Gin and Vodka in the summer.

Surprisingly, Alex confirmed they have no plans to make whisky as of yet, though he did confirm that they are continuing to experiment with their range and hope to put out some small-batch releases under the Glenrinnes Distillery brand.

Should you be intrigued as to how they make their spirits, the family-run team actively encourage visitors. With

11 different botanicals including cowberries and sorrel foraged from the Glenrinnes Estate

PREFERRED SERVE
Gin Martini. 60ml of Eight Lands Organic Speyside Gin and 10ml of quality white vermouth. Stir down over ice, or shake if you prefer. Strain into a chilled glass

PREFERRED GARNISH
Twist of lemon or grapefruit peel

RECOMMENDED COCKTAIL
EIGHT LEAF CLOVER

Ingredients
50ml Eight Lands Organic Speyside Gin
20ml Discarded Vermouth
10ml sugar syrup
15ml fresh lemon juice
4 or 5 fresh raspberries
20ml egg white
Angostura bitters to garnish

Method
1. Shake all ingredients over ice.
2. Double strain into a chilled glass.
3. Add three drops of bitters to garnish.

a range of tastings and tours, people can check out their bespoke 1,000-litre copper pot still Rebecca and their two towering 10-plate columns.

EIGHT LANDS GIN

Seeking to create a contemporary classic gin, using only 100% organic ingredients with no GM crops or pesticides, the family-run distillery is proud to be organic certified through SOPA – the Scottish Organic Producers Association.

Combining their top-quality Speyside spring water, drawn from the Clashmarloch spring on the lowest slopes of Ben Rinnes, with botanicals found on their estate including cowberry and wood sorrel, they've created a London Dry gin that balances the citric flavours of the sorrel, and the tart red berry notes of the cowberries, with the slight spice forest notes of the juniper.

Best of all, Alex recommends pairing the Eight Lands gin with their vodka to make an amazing vodka Martini – with the two spirits combining with vermouth to create a top-quality Vesper.

EL:GIN

WORKING IN THE whisky industry and being surrounded by many of the world's most famous Scotch distilleries, gave Leah Miller the desire to make her own mark on the world of Scottish based spirits.

Drawing upon her experience as a distillery manager and the desire to create a gin that would appeal to non-gin drinkers, she began the journey that would see her go on to create an exciting new spirit, named for the picturesque Moray town in which they are based.

Having a clear idea of the style they wanted, and spotting a gap for white spirits in the local area, Leah spent a year or so experimenting on a tiny 5-litre still before settling on a recipe that was the opposite of a 'ginny gin'.

However, something was still missing from the resulting sweet and easy-drinking gin that Leah couldn't put her finger on.

It was only the inspirational thought of adding oats to the mix that led them to the surprise discovery that this uncon-

BOTANICALS
Juniper, strawberry, raspberry, apple, coriander, angelica root orange peel and orris root

PREFERRED SERVE
Straight over ice

PREFERRED GARNISH
Raspberries and a slice of red apple for a sweeter serve or green apple for a crispier serve

ventional botanical seemed to bring all the flavours together and add a creamier, mouthfeel.

Launching in May 2016, their new signature gin is distilled, labelled and bottled by the team within the Findrassie Estate in Elgin; a couple of gin liqueurs, the Moray Berry and Moray Mocha followed just a short time later.

The Speyside Heather Honey Gin,

SMALL BATCH DISTILLED
SCOTTISH GIN

A gin from the heart of Malt Whisky
country, hand crafted in a traditional
copper still. Juniper is in here, but does
not dominate. We have created an expert
infusion of locally grown fruits with a
unique combination of Scottish
oats and botanicals.

40% vol   70cl

Findrassie, Moray
elgin-gin.co.uk
Price £35 / Quantity 70cl / ABV 40%

DOODLE PUP

Ingredients
25ml El:Gin
25ml Moray Berry
200ml apple and mango juice
Dash of lime
Lemonade to top up
Slices of lime, to garnish

Method
1. Add the El:Gin, Moray Berry and apple and mango juice to a mixer with ice.
2. Add the dash of lime, shake well then strain into a glass over ice.
3. Top up with the lemonade and garnish with the slices of lime.

which features local heather honey collected from hives located on the nearby Ben Rinnes, and a commemorative gin for the town's hospital, the Dr Grey Cherry Gin, were also added to their range.

The year 2020 was a big one for El:Gin, with the news that not one, but two of their gins, the Speyside Heather Honey and the Moray Berry had won gold at the Scottish Gin Awards. The exciting newest gin, Small Batch Citrus Gin, was released in 2020 and features two types of orange, pineapple and a touch of cardamom.

EL:GIN SIGNATURE

The base spirit for all of their gins, the El:Gin Signature is made using botanicals foraged from the beautiful Moray region by Leah and distiller Kelly Jack and comes in a striking black stonewear bottle.

Designed to be a smooth and easy-drinking gin, the pair use local botanicals like apples, strawberries and raspberries – as well as the all-important oats – to create a delicious, exhilarating spirit that has a wonderfully rich and luxurious mouthfeel that's as enjoyable neat as it is in a wee G&T.

Benromach Distillery, Forres
reddoorgin.com
Price £30 / Quantity 70cl / ABV 45%

RED DOOR

IT'S PERHAPS NO surprise that the hometown of noted botanist Hugh Falconer would provide the perfect setting for gin making. However, it is surprising that Forres, which is home to two whisky distilleries in the form of Dallas Dhu (sadly now silent) and Benromach, would take so long to have its very own gin.

Located close to the River Findhorn and the Romach Hills, Benromach Distillery is one of the Speyside region's smallest. Built in 1898, the whisky production site was mothballed in 1983 before being acquired by single malt whisky specialists Gordon & MacPhail in 1993, who set about rebuilding the distillery's name, before eventually re-opening in 1998.

Now a hugely popular Scotch brand in its own right – having recently celebrated its 20th anniversary – the team at the family-owned distillery decided that instead of basking in the success of their massively popular single malt whisky, they'd extend their spirits range to create a gin that reflected the wonderful scenery that surrounds the town. They spent months working with leading experts in the gin industry to create a gin recipe that would not only appeal to Scottish drinkers but – much like their whisky – would also appeal to drinkers all around the world, ultimately settling on a balanced mix of local and traditional botanicals.

Named after the distinctive red doors at Benromach Distillery, Red Door Gin was launched in July 2018, following the installation of Peggy, a handmade copper still – which is virtually a miniature replica of their spirit still – in the old malt barn at the distillery. Using a vapour-infused distilling method, and eight botanicals chosen specifically to capture the spirit of Scotland's 'majestic mountains, forests and wind-swept coastline', the London Dry gin is designed to be as at home in a personal collection as it is on the gantry in your best-loved gin bar.

Designed to reflect its eponymous red door, the gin comes in a ruby bottle

Juniper, pearls of heather, coriander seed, sea buckthorn, rowan berry, angelica root, lemon and bitter orange

PREFERRED SERVE
A chilled balloon glass filled with fresh ice and 1 part Red Door Gin to 2 parts good quality tonic

PREFERRED GARNISH
Fresh raspberries or a grapefruit slice

RECOMMENDED COCKTAIL
RED DOOR NEGRONI

Ingredients
25ml Red Door Highland Gin
25ml Campari
25ml good quality sweet (red) vermouth
Slice of orange to garnish

Method
1. Take a chilled 'Old Fashioned' glass – a tumbler or whisky glass is perfect.
2. Take a separate mixing glass with some ice.
3. Stir the ingredients in the mixing glass and add a little more ice.
4. Strain and pour over ice in your original 'Old Fashioned' glass and garnish with an orange slice.

and decorative box with its own sliding door. Also, keep an eye out for images of Scoobie, the distillery's pet cat and mouser. As a popular tourist attraction, those who visit Benromach to find out more about how this new handcrafted gin is made will soon be able to enjoy a new visitor experience that narrates the story of Red Door Gin and the engaging personalities who create it.

Not ones to rest on their laurels, they've just launched a new limited edition seasonal range, with the first release, Red Door with summer botanicals being released just in time for the summer.

RED DOOR HIGHLAND GIN
A small-batch London Dry gin, Red Door features a strong undercurrent of juniper punctuated with the sweet citrus of the bitter orange and lemon, while the fruity and floral notes of the sea buckthorn and pearls of heather battle it out with the deeper flavours of the rowan berries.

Bottled at 45% ABV, it doesn't betray its stronger alcohol content and works nicely in classic cocktails. It could just as well be enjoyed on its own with a few chunky ice cubes and some tart, fresh Scottish raspberries. We'd expect nothing less from one of the country's most popular up-and-coming spirit producers.

GIN NO. 5

USING HIS EXPERIENCE from decades working in the whisky industry in Speyside, distiller Duncan Morrison decided to give making the other love-of-his-life spirit a try. After chatting with his wife Shirley, the pair bought a small still for Christmas one year and began experimenting with recipes. Eighteen attempts later they settled on Batch No.5 as the one they loved.

Scaling up with a new 70-litre still, named Maddie after their granddaughter, they created their Roehill Springs Distillery in a building on their family-run farm in Keith, with the distillation, bottling, labelling and waxing all being done by hand on the farm.

Named for the spring in which they draw their all-important water, the Roehill Springs Batch No. 5 combines a blend of traditional and local botanicals (such as hand-picked rhubarb and elderflower). Duncan's distilling experience has helped create a truly exciting gin that's already gaining recognition with several top awards.

BOTANICALS
Juniper, coriander, orris root, angelica, pink peppercorns, cassia bark, lemon peel, rhubarb

PREFERRED SERVE
Add several large ice cubes to a balloon glass, pour Gin No 5 over the ice, add a slice of orange, top off to taste with Schweppes Indian Tonic Water and let the gin shine through

PREFERRED GARNISH
Slice of orange

Roehill Springs Distillery Limited, Moray
roehillsprings.com
Price £35 / Quantity 70cl / ABV 43%

GOODWILL

BOTANICALS
Juniper, orange peel, lemon peel, almond powder, orris root, hawthorn berries, cinnamon stick, coriander seeds and angelica root

PREFERRED SERVE
Over ice, with Fever-Tree Premium Tonic Water (or if you're feeling bold, Fever-Tree Elderflower Tonic Water)

PREFERRED GARNISH
Toasted orange peel and bruised coriander leaves

GlenWyvis Distillery, Dingwall
glenwyvis.com
Price £36.99 / Quantity 70cl / ABV 40%

SITTING PROUDLY UPON a hill overlooking Dingwall and the Cromarty Firth, with the stunning backdrop of the Ben Wyvis mountainside, GlenWyvis is Scotland's first community-owned distillery, with over 3,600 owners, all of whom put money towards its creation.

A community-benefit society, which means a share of all future profits will be invested back into community projects both locally and further afield, people from 37 countries have invested over £3 million to join the fledgling distillery group on their journey.

Dingwall's last whisky distillery closed in 1926, and, almost a century later, its successor, the new GlenWyvis spirits production site, began producing its own whisky spirit on Burns Night 2018. Since then, the distillery has produced over 850 casks and will be releasing their first single malt whisky at the end of 2021.

The distillery, keen to push sustainability, takes water straight from Ben Wyvis into the on-site borehole and uses renewable energy sources such as wind, hydro, solar and biomass.

Previously, their Highland gin had been made at Saxa Vord Distillery in Shetland however, the team installed their new gin still – affectionately named Heather – in summer 2018, bringing the production process home and creating this new gin in the process.

Made using nine botanicals, including spices, fruits and locally picked hawthorn berries, GoodWill Gin is the first major release from this brilliant little craft distillery. Since launch, the team have also released a selection of cask-aged gins – keeping a tie to their roots as a whisky-producing distillery.

GLENWYVIS GOODWILL GIN

'Made to share', GoodWill Gin comes in a fantastic bottle with a low curved neck and striking purple label which is embossed with icons relating to its creation. GoodWill has a crisp, full-bodied taste with distinct hints of citrus from the orange and lemon, followed by the spiced warmth from the coriander and cinnamon.

It's an invigorating gin that not only tastes great, but also helps to provide money towards strengthening and improving the Highland community. You can feel good about yourself while drinking GoodWill.

RECOMMENDED COCKTAIL

GLENWYVIS

Ingredients
50ml GoodWill Gin
25ml lemon juice
10ml Benedictine D.O.M.
10ml orgeat syrup
10ml ginger syrup
1 tsp orange marmalade

Method
1. Shake all the ingredients in a shaker until chilled (approximately 15 seconds).
2. Double-strain over crushed ice.

73
LOCH NESS

BOTANICALS
A closely guarded secret

PREFERRED SERVE
Loch Ness Spirits have three options
to suit non-tonic drinkers and tonic
drinkers alike:
Premium tonic with a slice of kiwi fruit
Ginger beer with fresh rhubarb
Soda water with a split vanilla pod

PREFERRED GARNISH
Kiwi fruit, rhubarb or vanilla pod

Loch Ness Spirits Ltd, Inverness
lochnessgin.co.uk
Price £45 / Quantity 70cl / ABV 43.3%

LYING JUST SOUTH of Inverness on the north-eastern bank of Scotland's most famous loch is a spirits company that aims to create a gin that will become as notorious as the monster that haunts the waters.

Loch Ness Gin was launched in August 2016 by Kevin Cameron-Ross and his wife, local GP Lorien Cameron-Ross, whose family has lived in the area for over 500 years. Lorien speaks of how the bones of her ancestors are 'literally lying in the ground' that grows their botanicals.

Talking of which, the pair keep the botanicals they use as a closely guarded secret. Though they enjoy affecting an air of mystery very much in keeping with the loch that gives their gin its name, Lorien adds that the real reason for the secrecy is to ensure customers can make their own judgements as to the taste profile. Even with this reluctance to disclose their recipe, the Cameron-Rosses are happy to explain that all the botanicals they use are sourced close to

RECOMMENDED COCKTAIL

NESS 75

Ingredients
20ml Loch Ness Gin
2 spoons spiced apple jelly
15ml fresh lemon juice
Champagne

Method
1. Pour the gin, jelly and lemon juice into a shaker, shake for several minutes.
2. Strain into a flute and top with Champagne.

the distillery, hand-picked on the banks of Loch Ness.

These include the abundant 'black gold' juniper, which they gather from their own crop. It's the beauty of this which led them to create their own gin.

The entrepreneurial pair credit the advice and guidance given to them by fellow Scottish distillers for the fact that their gin so quickly grew into a viable product that now flies off the shelves. Loch Ness Gin is made in four annual batches of just 500 at a time, owing to the availability not only of their rare native crop of juniper, but also other local botanicals sustainably sourced from the area.

The gin is then hand-bottled and hand-labelled on site. Since 2016, they have added a new gin to their range, Loch Ness Legends, and the world's first absinthe, Loch Ness Absinthe, to contain Scottish wormwood.

LOCH NESS GIN

Hand-distilled in a 50-litre still, Loch Ness Gin has won multiple awards since its release. Produced in small batches, it's lauded as a premium gin with 'perfumed floral aromas combined with delightful fruity notes and just a hint of pine'.

The packaging is as intriguing as the gin itself, with a semi-translucent black bottle, featuring hot-foiled copper, and a stylised illustration of the loch's legendary monster on the label. The ABV also has a very specific story behind it: the 43.3% the spirit is bottled at reflects the average depth of the loch – 433 feet.

BOTANICALS
Juniper, orange peel, angelica root, cloudberry, Scots pine, coriander seed, kaffir lime leaves, elderflower, orris root

PREFERRED SERVE
Can be enjoyed neat over ice or paired with premium tonic and a slice of green apple

PREFERRED GARNISH
Raspberries and blackberries

Cairngorm Gin Company Limited, Carrbridge
cairngormgin.com
Price £39 / Quantity 70cl / ABV 43%

74
CAIRNGORM GIN

HAVING GROWN UP surrounded by one of Scotland's most spectacular national park, Jack Smith combined his love for the outdoors and the native flora and fauna of the Cairngorms with a passion for hospitality, to create a compelling new Highlands gin.

With the help of experienced distilling guru Lewis Scothern at Distillutions, Jack used his knowledge of cooking, gained from working with a number of prominent Michelin star chefs, and the desire to use only the best ingredients, to explore the abundance of natural botanicals to be found in the area to set about making his new Cairngorm Gin.

This new premium gin aimed to capture the essence of the stunning Highland scenery that surrounds his newly-established micro-distillery.

The rare cloudberry, which grows in some remote areas of the Cairngorms, was chosen as the key element of a recipe that follows the classic London Dry formula but has Jack's own flavoursome twist.

CAIRNGORM ESPRESSO MARTINI

Ingredients
60ml Cairngorm Gin
60ml Kahlua
60ml espresso
A few coffee beans

Method
1. Add ingredients to a mixer with ice, shake then strain into a martini glass.
2. Add the coffee beans to garnish.

Jack has named their copper pot still Ginger after his late grandfather, a bit of a character who was known for saying it was easier to make a G&T than 'a cup of tea'. It's now become a tradition for the team to toast 'Ginger' with a gin from each batch they distil.

CAIRNGORM GIN

Produced in small batches, of no more than 70 bottles, Cairngorm Gin is a culmination of a process that sees Jack combine the all-important cloudberries with eight other botanicals including Caledonian Pine, elderflower and water collected by hand from the River Spey.

Only using the heart of the cut, with the heads and tails discarded from each run, and allowing the final spirit to rest for a week after it comes off the still, leads to the creation of a gin that really allows both the juniper and sweet, tart notes of the cloudberry to shine.

NORTHERN SCOTLAND

Pixel Spirits Ltd, Fort William
pixelspiritsltd.co.uk
Price £39.90 / Quantity 70cl / ABV 42%

DEVIL'S STAIRCASE

NOT CONTENT WITH buying and running a small hotel bar and restaurant in what is some of the most breath-taking scenery the country has to offer – the Western Highlands and the Grampian Mountains – Craig and Noru Innes decided to build their own gin distillery.

Loch Leven Hotel, which sits on the northern shore of Loch Leven, is the home to Pixel Spirits and the location of one of the country's most unique craft distilleries and gin schools.

The journey towards gin production began for the pair when they decided to commission a branded gin for the bar. Having contacted a few contract distillers to no avail, they decided to simply do it themselves.

The construction of the new distillery began in an almost derelict seventeenth-century A-frame barn – part of croft buildings currently on the grounds of Loch Leven Hotel – in 2015. Craig, the master distiller, along with another member of the hotel's staff, who

happened to be a carpenter, completed the build in October 2017. All the work was done in-house with only the electrics and more complicated plumbing jobs outsourced to local tradespeople.

The first batch of Devil's Staircase,

named for the ominous, outrageously steep off-road track that's part of the West Highland Way, was created just a few months later. Made in small batches, Devil's Staircase is handcrafted on their still, affectionately named Orsetta – Italian for 'little bear' – from start to finish, with each bottle hand-signed and numbered by Craig.

The husband-and-wife team were passionate about introducing a further range of gins and spirits as the business expanded and launched Drookit Piper, a citrus forward expression, in 2019. This was followed by their extremely limited batch Artisan Range in 2020 with speciality products such as Chanterelle Gin and a small batch white rum.

They share this passion for gin at their school, where fellow enthusiasts can learn the art of distilling on traditional copper pot mini-stills and create their own gin recipes. Located within the historic former byre of the distillery, the school is open to gin fans from absolute beginner to expert, with each guest taking home a bottle of their creation upon completion of the four-hour experience.

DEVIL'S STAIRCASE GIN

Craig describes the Devil's Staircase as the 'perfect blend of warm spice and citrus zest' with ten botanicals used to create the refreshingly sweet (juniper, coriander, orange peel and lemon peel) and spicy (the grains of paradise, cardamom, nutmeg and cassia) profile of this fun Highland spiced gin.

The label, created by the renowned Scottish illustrator Iain McIntosh, is a work of art itself, featuring a fun illustration of a red devil beginning his ascent of a stone staircase, reflecting the rugged trek that gives the gin its name.

THE SPICED BEE'S KNEES

This is the producer's take on a Prohibition-era classic – the bee's knees. It's the perfect summer serve: the spice from the gin and sweetness of the honey balanced with the acidity of the citrus make it the ultimate refreshing cocktail.

Ingredients
50ml Devil's Staircase Gin
2 tsps Ed's Bees Glasgow Honey (or other pure honey)
20ml freshly squeezed lemon juice
20ml freshly squeezed orange juice
Orange peel, to garnish
2 to 3 cardamom pods, to garnish

Method
1. Chill a cocktail glass.
2. In the bottom of a cocktail shaker, add the gin and honey and stir until the honey has dissolved, then add the freshly squeezed orange and lemon juices and shake over ice for 20 seconds to chill the drink.
3. Discard the ice from the glass and strain the cocktail straight into the glass.
4. Garnish with an orange peel and a couple of cardamom pods to serve.

NORTHERN SCOTLAND

BOTANICALS
Angelica seeds, kaffir lime leaves, lime peel, orange peel, allspice, juniper

PREFERRED SERVE
Ideal with ice cubes and a decent tonic

PREFERRED GARNISH
The Fairytale team say this gin doesn't need a garnish

Fairytale Distillery, Ardelve
fairytaledistillery.co.uk
Price £43 / Quantity 70cl / ABV 41%

SITTING CLOSE TO the shores of Loch Alsh in the western Highlands is what might just be one of the UK's most magical gin distilleries.

Founded in 2018 by German couple Thomas Gottwald and Manuela Kohne-Gottwald, who moved to the Ardelve area in 2013, the Fairytale Distillery is a tourist attraction in its own right and will be unlike anything you'll have seen before.

Looking like it's been lifted directly from the pages of *Hansel and Gretel*, the whimsical building has crooked chimneys, slanted windows and a sweeping slate roof that almost touches the ground on either side of striking pea-green walls.

Thomas originally came up with the idea to make gin, after becoming friends with gin lover Roger Knight, who was in turn introduced to gin by his uncle who was chief steward on the ocean liner *Queen Elizabeth II*. The pair then put together a plan that resulted in them using water from the spring-fed burn Alt Mor na Dornie and enrolling botanists Martin Wünsche and Stefan Lipka to help create their new range.

Seeming to float above its own little lochan, the distillery, which houses a little 80-litre Austrian copper still named Tinkerbelle, now produces an extensive range of interesting gins that come in black stone bottles which are numbered rather than named.

They now have no less than four London Dry Gins, four Western Style Gins, four Navy Strength Gins, one Old Tom and a Sloe Gin, as well as a rum and even an absinthe.

This multi-award-winning gin is Fairytale's best-seller and is a full-bodied London Dry as magical as the distillery that makes it, with rolling flavours of juniper, citrus and those lovely floral notes.

MAGIC GARDEN

Ingredients
50ml Highland Gin No. 6
25ml elderflower liqueur
75ml apple juice
15ml fresh lime juice
Cucumber slices, to garnish

Method
1. Add all the ingredients to a mixer with ice.
2. Shake then strain into a tall ice filled tumbler.
3. Add the cucumber slices to garnish.

View of Beinn-na-Cailich, Isle of Skye, from Ardelve

NORTHERN SCOTLAND

Dornoch Distillery, Dornoch
thompsonbrosdistillers.com
Price £32.50 / Quantity 70cl / ABV 45.7%

77

THOMPSON BROS

THERE CAN BE few grander settings for a gin distillery than a fifteenth-century Highland castle that is already home to more than a few spirits. Set within the historic seaside town of Dornoch, which lies across the Dornoch Firth from Tain in the heart of North Coast 500 country, Dornoch Castle Hotel is a former bishop's palace turned castle turned hotel.

Owned and run by the Thompson family since June 2000, what was once simply a castle 'haunted' by a sheep rustler, is now more famous for the liquid spirits that are poured freely in its hugely popular and award-winning whisky bar.

It was in 2015 that brothers Phil and Simon Thompson decided to take the expertise and passion they'd built up behind the bar and follow their dream of opening their own distillery.

Using a one-storey stone shed on the castle grounds that had previously housed a fire station for the local community, the pair launched a crowd-funding drive and eventually installed the versa-

tile distilling equipment which includes a mash tun, wooden washbacks and several copper stills, before producing their first spirit in 2017.

Following lots of experiments with recipes and ratios, gathering essential feedback from the people who helped fund their distillery, the brothers settled on a formula for their Organic Highland Gin.

BOTANICALS
Juniper, coriander seed, black peppercorn, dried lime, dried lemon, angelica, cardamom, meadowsweet, lemon peel, orange peel, bergamot, thyme, rosemary and bay leaf

PREFERRED SERVE
Fresh lemon peel served with Fever-Tree Original Tonic Water, though the team say it works just as well neat over ice

PREFERRED GARNISH
Fresh lemon peel

NORTHERN SCOTLAND

ROYAL HAWAIIAN

Ingredients
40ml Thompson Bros. Organic
  Mediterranean Gin
15ml orgeat syrup
10ml lemon juice
25ml fresh pineapple juice

Method
1. Pour all the ingredients into a cocktail shaker with ice.
2. Shake for 30 seconds.
3. Strain into a chilled coupe glass.

Uniquely, they use a small percentage of Plumage Archer barley, which is floor-malted, organic and the 'first genetically true barley variety', to make their gin.

This links it to the production of their very own single malt, which they have just released the first batches of.

They are also keen to explore the process of how they make their spirits.

They say that the long distillation of 36 hours is key to ensuring every botanical is given time to bed in, helping to create the fullest flavour-journey possible. This patience and care extends to the dilution process with the brothers taking time to add the water gradually so it doesn't have any adverse effect on the spirit.

Having already reached capacity at their current distilling site, Dornoch is set to expand and have launched another crowd-funder in a bid to create a bigger distillery as well as facilities for tourists who wish to come and see the process in person.

The move will entail an expansion of production capacity, a new retail space and tasting room and new jobs for the local Dornoch community.

THOMPSON BROS. ORGANIC MEDITERRANEAN GIN

Launched in November 2017, Thompson Bros. Organic Highland Gin has an excellent depth and mouthfeel – thanks to the body provided by 10% of the base being derived from the narrowest cut of their new-make malt spirit.

Their latest edition, the Organic Mediterranean Gin, which replaces their first version, builds on the original recipe and adds single distillations of fresh seasonal organic citrus fruits Bergamot, Lemon and Orange, and the herbs Rosemary, Thyme and Bay Leaf.

Slowly distilled and blended, it's a bright, clean and citrus led gin that will remind you of the region that gives it its name.

The striking artwork that features on each of the bottles has been provided by London-based artist and illustrator Karen Mabon.

78
CROSSKIRK BAY GIN

BORN FROM A shared love of spirits, the plans to create North Point Distillery sprung from founders Alex MacDonald and Struan Mackie's passion for whisky, rum and gin that saw them become fast friends when they first met five years ago in London's fast-growth tech start-up scene.

After Struan moved back to his home country of Caithness on the Northern Coast in 2020, the two came to the decision to actually start making spirits instead of just drinking them.

Though the country was gripped by the pandemic, they still managed to acquire their ideal site located in the Forss Energy and Business Park just outside of Thurso. Following this up by hiring their first employee, Laura Mackenzie, they started work on-site in August 2020.

And if that's not all, even with the challenges they faced in what was an incredibly tough year, they still managed to create three remarkable award-winning spirits, with two expertly made

BOTANICALS
Juniper, angelica, Icelandic moss, coriander, heather, cassia, rowan berries, cinnamon, lemon verbena, cardamom, sea buckthorn, orris, grains of paradise, Scots pine, grapefruit peel, lemon peel, orange peel

PREFERRED SERVE
Indian tonic and ice

PREFERRED GARNISH
Twist of pink grapefruit

North Point Distillery, Thurso
northpointdistillery.com
Price £35 / Quantity 70cl / ABV 45.1%

rums – the Pilot Rum and the Spiced – joining their award-winning Crosskirk Bay Gin.

Reflecting the beautiful and rugged Highland bay on which it is made, this gin is made using locally sourced botanicals including Icelandic moss, rowan berries, and heather combined with more traditional botanicals sourced from Fairtrade and environmentally friendly suppliers.

Designed to be a local, sustainable, London Dry juniper spirit, the team use a one-of-a-kind still, aptly named Sandy Stroma, after Alex's mother Sandie and Struan's father Sandy, and Stroma which is the abandoned island off the coast of Pentland Firth.

The team use in-pot maceration, which in turn creates a punchy and flavoursome gin, with a refreshing citrus character. Interestingly, one of the main botanicals, which is close to their hearts, is Scots pine which is taken from Struan's first family Christmas tree, planted in their garden way back in 1993.

CROSSKIRK 75

Ingredients
50ml Crosskirk Bay Gin
1 tbsp lemon juice
1 tbsp sugar syrup
Ice
Champagne and lemon zest, to top up

Method
1. Pour the lemon juice, sugar syrup and gin into a cocktail shaker then fill up with ice, shake.
2. Strain into a cocktail glass, top up with Champagne and top with lemon zest.

BADACHRO

THE TALE OF Badachro is one of serendipity as much as it is one of provenance, and of the lasting relationships between people and place. An inn in a former fishing village close to the spectacular Loch Gairloch in the Northwest Highlands was the scene of the first meeting between Gordon and Vanessa Quinn.

The couple went on to marry and then, eventually, by way of the Middle East and London, set up a bed and breakfast that would form the beginnings of their distillery. Tired of the rat race, Gordon and horticulturalist wife Vanessa returned to Badachro to build themselves a new home and a life in the area where they first met.

In a desire to create a keepsake for the tourists who visited their bed and breakfast, the pair decided to create a gin that was 'truly of Badachro', something for guests to take home as an authentic expression of the place. Vanessa, who Gordon describes as the more creative of the pair, turned an old joke about

Badachro being so special 'because of the smells' on its head and that became their inspiration.

Instead she pointed out that the region is actually filled with lovely smells such as the sea breeze, the gorse blossoms in the summer, and the wild myrtle after the rain. The couple decided to take this idea forward, thinking that if they could bottle these wonderful smells they could create their keepsake. First, they experimented with a myrtle liqueur – Mirto, a drink popular in Italy – then a flavoured vodka, before settling on what would become their gin, which they launched in spring 2017.

A family affair, the distilling is presided over by Gordon while Vanessa leads the foraging and takes care of recipe development and quality standards, which the children, Sean and Ashley, help out with wherever possible.

One of the most industrious producers you'll find, they've since added a Coastal Gin – made with botanicals foraged around the coast of Gairloch and

Badachro Distillery, Wester Ross
badachrodistillery.com
Price £36.95 / Quantity 70cl / ABV 42.2%

Juniper, myrtle, elderflower, lavender, rose hip, gorse blossom, liquorice roots, orris roots, angelica, citrus, lemongrass and lime leaves

PREFERRED SERVE

Badachro is gaining quite a reputation as a sipping gin. It's perfect to enjoy with tonic and a garnish of a slice of lime and cardamom pod – not crushed or bruised – which adds a pleasing dryness to the finish

PREFERRED GARNISH

A slice of lime and a cardamom pod

RECOMMENDED COCKTAIL
BADACHRO BRAMBLE

The Quinn family make their own cassis and use it to create their distinctive version of a bramble.

Ingredients
50ml Badachro Gin
25ml freshly squeezed lemon or lime juice
5ml sugar syrup
25ml crème de cassis (Badachro's own or otherwise)

Method
1. Mix it all together in a shaker and pour over ice.

Badachro – the Storm Strength Gin, and three full strength flavoured versions in the form of the Orange Orains, the Raspberry and the Sloe Gins.

The hard-working team also have a super-smooth, characterful Dancing Puffin vodka and their popular Bad na h-Achlaise single malt whisky range as well as running two stunning self-catering accommodations.

BADACHRO GIN

The couple hand-picks the botanicals such as myrtle, elderflower, lavender, rose hip and gorse blossom from within 500 metres of the distillery but, as Gordon explains, while they would love to pick their own juniper it doesn't grow abundantly on the west coast because of the climate.

Instead, they source most of it from Croatia and Italy and combine it with the hand-picked plants and traditional gin botanicals such as liquorice roots, orris roots, angelica, citrus, lemongrass and lime leaves to create their gin. The lime leaves were added after experiments with citrus peels resulted in slight hints of bitterness. The resulting gin is not your typical flat-profiled London Dry gin.

It's full-flavoured, quite fruity to start with, spicy and sweeter in the middle with the juniper and myrtle and, at the end, there are citrus notes with the possibility of picking up on the lavender.

CROFTER'S TEARS

BOTANICALS
Juniper, coriander, cardamom, orange peel, cubeb, cassia, orris root, lime peel, purple heather flowers

PREFERRED SERVE
Fever-Tree Mediterranean tonic with a twist of lime and a bramble

PREFERRED GARNISH
Twist of lime

Newblack Croft, Smerral
iceandfiredistillery.com
Price £38 / Quantity 70cl / ABV 40%

BORN FROM A desire to sustain the crofting lives of its founding members, Ice & Fire Distillery was set up to supplement the livelihoods of a family of Highland crofters.

Husband-and-wife team Stephen Wright and Jaqueline Black decided to make spirits from the local water and botanicals, after Jaqueline's brother became ill and could no longer keep up with his role as a gamekeeper.

The trio then set about creating a distillery on their Highland croft in Smerral, Latheron, Caithness that would hark back to the days when many crofters in this stunning Highland region would supplement their meagre incomes by distilling illicit whisky.

The hills and moors of Caithness have a rich history in this thrilling act of lawless spirit making with a ready supply of water from the burns, peat from the hills and grain from the land, ensuring they had all they needed (along with a little dose of cunning and luck to evade the excisemen) to produce the 'peat-reek' – one of Scotland's earliest forms of the uisge beatha.

Using handcrafted copper stills, they now produce their Caithness Highland Gin, made using rhubarb – still a staple crofting ingredient – and their Crofter's Tears Gin, made using purple heather – which blooms beautifully across the hills in summertime.

The enterprising team are not just sticking to gin, with the Caithness Raiders Rum and a new dark rum recently added to their line-up.

CROFTER'S TEARS GIN

With a nod to their Highland Crofting heritage, the bottle for this punchy gin features beautiful artwork of the Caithness scenery on the label, as well as purple heather, which is their signature botanical, etched directly onto the vessel itself.

Produced on their expert 200-litre Hoga Stills, this deliciously smooth London Dry gin is created using purple heather tips, which only bloom in July and August and are hand-picked by the team and blended with juniper and a range of botanicals including cassia bark and cubeb to produce a flavoursome spirit with a rich, oily, mouthfeel that will stand up to even the heaviest of mixers.

CROFTER'S CLOVER CLUB

Ingredients
50ml Crofter's Tears
1 tbsp dry vermouth
Blackberry syrup
25ml lemon juice
1 tbsp egg white
1 or 2 blackberries, to garnish

Method
1. Put all of the ingredients in a shaker, then shake hard for a minute or two.
2. Add ice then shake until the shaker feels cold, double strain in a cocktail glass and add a blackberry or two to garnish.

NORTHERN SCOTLAND

Dunnet Bay Distillers, Caithness
dunnetbaydistillers.co.uk
Price £34 / Quantity 70cl / ABV 41.5%

ROCK ROSE

THE UK'S MOST northerly mainland spirits distillery, Dunnet Bay distillery, was launched in 2014 after engineer Martin Murray decided to leave the oil and gas industry to return home to start up a new venture with his wife Claire.

Born of his passion for brewing and distilling, something he had studied as part of his master's degree in Chemical Engineering at Edinburgh's Heriot-Watt University, Martin joined Claire in Caithness in 2011 to plan the distillery they would build there and the spirits that would flow from its stills when it began production. George, the couple's much-loved dog, was involved in the development of the distillery and brand. He is even name-checked on the website as one of the company's founding partners.

Located in the spectacular bay of Dunnet, the distillery is home to their bespoke pot still, Elizabeth, which was designed especially for them and produces the 500-litre batches of

BOTANICALS
Juniper, sea buckthorn, Rhodiola rosea, rowan berries, blaeberries, cardamom, coriander seed and verbena, to name but a few

PREFERRED SERVE
Serve with lots of ice in a balloon glass with Fever-Tree Tonic and a curl of orange

PREFERRED GARNISH
A curl of orange or, if you want to try something a little different, garnish with a sprig of rosemary, toasted for full effect

their popular Rock Rose Gin. Every bottle of which is filled, hand-waxed, batch-numbered and signed before it leaves their distillery. The small team is firmly rooted in the community, with Caithness residents credited as being

NORTHERN SCOTLAND

brand ambassadors, and family members working at the distillery.

The brand is renowned for their gin range, which includes the original Rock Rose, a Navy Strength and Distillers' Edition and several limited edition seasonal expressions. In addition, they have launched a vodka named Holy Grass – known as bison grass in Poland – for the unusual botanical that was once found on the banks of the River Thurso and which gives it its sweet flavour. The distillery offers tours (best book in advance) and are proud of their green credentials, as thanks to a new solar PV system they now produce more power than they use.

They were also the first Scottish distillery to offer a freepost return, fully recyclable pouch for refilling their collectable Rock Rose Gin ceramic bottle.

ROCK ROSE GIN

Named after the 'Rhodiola rosea', the Rock Rose, the original gin released by the distillery, uses the intriguing botanical that grows on the cliffs of Caithness.

The Rhodiola rosea is famed for its health benefits. It's said that the Vikings once sought it out – thanks to its strength and vitality-giving qualities – and it seems that this fame has lasted into more contemporary times. The first batch of the gin sold out in less than 48 hours – an industry record. Interestingly, Martin uses juniper berries from two different countries, Italy and Bulgaria, and both varieties offer specific individual qualities to the final gin.

The recipe for Rock Rose was a long time in the making. Distiller Martin took eighteen months, eighty botanicals and fifty-five experiments to perfect it; the final version uses no less than eighteen botanicals, five of which are grown locally, including sea buckthorn and rowan berries. But even with all these immaculate details, one of the biggest selling points for Rock Rose is its white ceramic bottle. With its hand-waxed seal and gentle patterns, it stands out on any gin shelf.

RECOMMENDED COCKTAIL
PEPPERMAN

Ingredients
50ml Rock Rose Gin
25ml grapefruit juice
12.5ml sugar syrup
Pinch of ground pink peppercorns
Soda
3 strawberries

Method
1. Take a cocktail shaker, fill the glass half of it with ice and then pour the gin, grapefruit juice and sugar syrup over the ice.
2. Add a pinch of ground pink peppercorns with 2 strawberries.
3. Shake for several minutes then fill a highball glass with ice.
4. Single-strain the liquor into the glass.
5. Top with soda water and garnish with a strawberry.

TOP TIPS
Keep your eyes peeled for special editions, which use unique botanicals to achieve seasonal flavour profiles; these include Christmas tree spruce tips for 2017's festive gin.

The Coach House, Tain
sutorsgin.com
Price £38.95 / Quantity 70cl / ABV 45.4%

82

SUTOR'S GIN

THE IDEA FOR this 100% Scottish gin sprung from a birthday present that founder Stuart Wells wanted to create for his gin-loving wife, Clair, in 2016.

After successfully creating the exciting gift with neighbour Ed Scaman, Stuart decided he wanted to have a shot at creating spirits full-time with the caveat that they had to use only 100% Scottish ingredients.

Setting about replacing some of the 'off the shelf' ingredients he had used in their original gift recipe, he embarked on a four-year journey with Ed to not only perfect the recipe but set about creating the spirit that would go along with it too.

Purchasing a Genio Still, which is named Karen after an ongoing joke, and converting a nineteenth-century coach house near Tain in the Highlands, the pair wanted to create a new gin that was unquestionably of the area in which it was made.

Desiring to make the whole product, Stuart and Ed experimented with malting processes, optimal grain bills and locally sourced botanicals to create a gin that featured sustainably sourced ingredients, Scottish grain spirit made by them and even Scottish juniper sustainably sourced in the Highlands.

Celebrating the farming heritage of their local area, Stuart sourced barley from a local farm just two miles from the distillery that they use to create the barley wheat spirit that had a rich mouthfeel and worked in perfect balance with the wonderful botanicals sourced from around the area – like sea buckthorn picked at Nigg Beach and Douglas fir foraged from the nearby forests.

As part of their core range, they also do a premium Scottish vodka and their most recent release, Flora, is a wheat-based gin distilled using native Scottish flowers such as elderflower, sweet cicely, hawthorn, and violets, some of which the team are propagating themselves to help the local ecology in the area.

Stuart confirmed they hope to add a bigger still in the future and as well as experimenting with different grains

Scottish juniper, sea buckthorn, bog myrtle and Douglas fir

PREFERRED SERVE
Over ice with premium tonic and a slice of grapefruit

PREFERRED GARNISH
A slice of grapefruit

RECOMMENDED COCKTAIL
AULD ALLIANCE
(Craft56.co.uk)

Ingredients
25ml Sutor's Gin
20ml elderflower cordial
20ml lemon juice
Prosecco or Champagne to top up

Method
1. Add the ingredients to a cocktail shaker and shake for a couple of minutes.
2. Strain into a Champagne flute and top up with prosecco and Champagne.

is looking to begin producing a new limited edition range known as Sutor's Elements – seasonal liqueurs and spirits including a sloe gin and a wild cherry and wild raspberry gin liqueur.

SUTOR'S GIN
Handcrafted from start to finish at the distillery, Sutor's Gin work with local farmers to develop and produce the ideal grain for their base spirit, which they combine with locally sourced botanicals from the mountains and coastal areas that surround their Highland home.

Described as a love letter to where they live, this Grain to Grass gin is characterful and rich, owing to the complex base spirit and Scottish botanicals like sea buckthorn, bog myrtle and Douglas fir.

The Scottish juniper they use has a milder flavour than that sourced in Macedonia or Italy, but still creates that traditional fresh piney flavour that leads on the nose before the citrus and spice come through on the finish.

SEVEN CROFTS

SPORTING ONE OF the most instantly recognisable (and stylish) bottles out there at the moment, Seven Crofts Gin was the flagship spirit released by the Ullapool-based Highland Liquor Co.

Founded by partners Helen Chalmers and Robert Hicks in 2018, within view of the stunning Loch Broom on Scotland's north-western coast, their exciting new gin range takes its name from the original settlers who lived in the first seven dwellings in the picturesque fishing village.

A destination in its own right, Ullapool is known for its stunning scenery, history as a key herring port, and perhaps most surprisingly as Scotland's pre-eminent geological hot spot.

Taking 18 months to develop, and 96 different recipes, before they were able to settle finally on their perfect gin, the pair enlisted head distiller Ben Thompson and bought two Portuguese-made copper stills, one named Ruth (after Ben's wife) and the other Ella (after

Helen and Robert's firstborn) to bring their exciting new spirit to the masses at the tail end of 2019.

An 'international gin with a local heart', the Seven Crofts team are closely knit with the community in which they are based, and Ben celebrates this by regularly creating gins that are only available to buy in their small distillery-based bottle shop.

A regular example of this being their Ullapool gin which harnesses the essence of the area's local flora to create a gin that's 'reminiscent of the heather-covered hills' surrounding the loch and a great way to take a little bit of Ullapool home with you.

Also, following their original classic London Dry Gin, was the launch of their second core gin, the Navy Strength bottling which paid homage to the village's history as a key fishing port, dubbed Fisherman's Strength.

Bottled at 57% ABV, the small-batch gin is hand-distilled and has a flavour profile that's traditional and juniper led

Highland Liquor Company, Ullapool
highlandliquorcompany.com
Price £40 / Quantity 70cl / ABV 43%

BOTANICALS
Juniper berries, coriander seed, angelica root, pink peppercorn, cubeb, cardamom, fresh lemon peel

PREFERRED SERVE
Serve in a straight glass with lots of ice and a good quality tonic, (1 part gin to 2 parts tonic)

PREFERRED GARNISH
A slice of orange

RECOMMENDED COCKTAIL
SOUTHSIDE SEVEN

Their in-house twist on the classic Southside.

Ingredients
25ml Seven Crofts Gin
25ml fresh lime juice
25ml sugar syrup
6 mint leaves

Method
1. Shake the ingredients in a shaker with ice.
2. Double strain and pour into a high-ball glass with ice.
3. Top with sparkling elderflower soda.

with seven botanicals including pink peppercorn, cubeb, fresh lemon peel, juniper and coriander seed.

SEVEN CROFTS GIN
Working with Glasgow based creative agency D8, the Highland Liquor Co. have created a lovely bottle that's reminiscent of Dutch genever bottles, with the green ombre designed to resemble the fading from farmland to the clear waters of Loch Broom, and the ridges on the bottle chosen to not only make the bottle easy to pour but also to represent the contours of the land surrounding the distillery.

The Seven Crofts London Dry Gin itself is a classic style gin, produced in small batches, and features seven key botanicals including juniper berries, coriander seed, angelica root, pink peppercorn, cubeb, cardamom and fresh lemon peel.

The resulting gin is as luxurious as the bottle it comes in, with the juniper really sitting at the fore with waves of citrus and spice balancing well on the palate.

THE GINS

THE ISLANDS
WESTERN ISLES, SHETLAND & ORKNEY

ISLE OF
CUMBRAE
DISTILLERS LTD
MILLPORT, SCOTLAND

NOSTALGIN
HANDCRAFTED IN MILLPORT

BOTANICALS
Lavender, bramble, orange

PREFERRED SERVE
Tonic and ice

PREFERRED GARNISH
A slice of orange

11 Guildford Street, Isle of Cumbrae
isleofcumbrae-distillers.com
Price £38.50 / Quantity 70cl / ABV 43%

NOSTALGIN

ORIGINALLY MEETING IN 2019 over a shared love of gin, the group of friends that would go on to become Isle of Cumbrae Distillers set out to create a spirit that celebrated the island they call home.

Julia Dempsey, Lynda Gill, Bronwyn Jenkins-Deas, Philippa Dalton and Jenine Ward, opened the doors to their new venture in September 2020.

This international all-female distilling team, which hails from the UK, Canada and the US respectively, first came together with an idea to help raise funds to save Millport's historic Town Hall building, a cause they say remains close to their hearts and will feature as part of a future release.

One of only three Scottish distilleries to successfully launch during the pandemic, their signature gin, Nostalgin, is so named due to the feeling of nostalgia many people have for the island town of Millport which, in its heyday in the 1950s and 1960s, was a favourite family holiday destination 'doon the watter' from Glasgow.

Their second gin, Croc Rock, celebrates Millport's painted Crocodile Rock, which has been a much-loved landmark on the island for over a hundred years.

Keen to share their story, they've also created a special tour experience that highlights the key role women played in the history of gin as well as exploring the colourful history of eighteenth century smuggling along the Clyde coast.

Environmentally conscious and aware of the impact plastic has on Cumbrae's beaches and wildlife, the quintet have committed to working in a socially responsible and environmentally friendly way.

Millport holds a strong emotional connection for many of those who have visited the island or spent their childhood summers there, with the aptly named Isle of Cumbrae Distillers signature gin, Nostalgin, celebrating those who still come back to the island, and now share that experience with their own children and grandchildren.

Featuring a beautiful label of the harbour drawn by talented young artist Jack Sutton, who specialises in ink drawings, Nostalgin is made using lavender, bramble and orange to recreate the feeling of relaxation and calm that people experience on the short eight-minute ferry ride to the island from the mainland town of Largs.

RECOMMENDED COCKTAIL

HUSH HUSH

Ingredients
37.5ml Nostalgin
20ml red vermouth
20ml lemon juice
20ml blackberry shrub*
500g blackberries
Sugar, a generous sprinkling

Method
1. Shake all of the ingredients in a cocktail shaker and strain over crushed ice in a highball glass.
2. Garnish with a blackberry, grapefruit slice and mint sprig.

*For the Blackberry Shrub:
1. Cover 2 punnets of blackberries (about 500g) in sugar. Leave covered at room temperature to naturally turn into a thick syrup.
2. Strain out the berries and add apple cider vinegar to the syrup (4 parts syrup to 1 part apple cider vinegar).

THE ISLANDS

85

ISLE OF BUTE

BOTANICALS
Scottish fresh oyster shell, juniper,
coriander, angelica, seaweed,
cucumber, lemon

PREFERRED SERVE
Try this unique gin neat first, though if
you do fancy making a G&T, serve over
ice with a 2:1 ratio of tonic to gin

PREFERRED GARNISH
In a G&T, it's recommended to use chilli,
ginger and coriander, however, any
garnishes that complement the uniquely
savoury, coastal flavours will work well

Rothesay, Isle of Bute
isleofbutegin.com
Price £42 / Quantity 70cl / ABV 43%

THE TINY BUT accessible Isle of Bute which lies off the west coast close to the Ayrshire town of Largs, has been a favourite of holidaymakers travelling 'doon the watter' from Glasgow for generations.

Unsurprisingly, not only does it serve up inspiration for these city-dwelling daytrippers, but also captured the heart of experienced distiller and brewer Simon Tardivel, who fell in love with the island while living there.

The story of this fledgling distillery begins with beer, as after a chance meeting with owner Rhona Madigan-Wheatley at a trade show for brewers in 2018, a discussion around the rise of craft gin led to the pair mulling over the idea of creating their own.

A reason for the pair to discover more about the island itself, the decision led them to sourcing botanicals such as gorse, which they hand-pick locally from Mount Stuart Estate, and, remarkably, oyster shells, which they sourced from the famous Loch Fyne in Argyll as they believe they perfectly reflect the beautiful waters that surround their island home.

Produced on their 200-litre copper pot still, Audrey (Simon is a huge Audrey Hepburn fan), in the heart of Rothesay, they released their first two gins, the Gorse and Oyster Gin, in early 2018.

Their Heather Gin, with its hand-picked wild purple flower, followed next before the Island Gin, with its bottle artwork painted by Scottish landscape artist Emma S Davis, and their nod to whisky, the Oaked Gin, followed soon after.

They recently opened their own Gin Garden at the back of the Mount Stuart mansion house where guests can enjoy G&Ts, gin-tasting flights and even a small selection of locally sourced snacks.

OYSTER GIN

The world's first Oyster Gin is made by charging the still with fresh oyster shells giving the resulting gin a fresh coastal flavour, with subtle hints of sea salt and brine that, when combined with the citrus of the lemon, creates a savoury gin that is not only ideal as a base for mixing but also tastes delicious neat.

BUTE SNAPPER

Ingredients
50ml Oyster Gin
75ml tomato juice
25ml lemon juice
20ml chilli and tomato shrub (can be omitted to make the recipe a bit easier!)
Tabasco sauce, Worcestershire sauce, cucumber, chilli, basil, thyme

Method
1. Add all ingredients to a cocktail shaker and roll between 2 cocktail shakers with cubed ice.
2. Fine strain into a highball glass over cubed ice. Garnish with cucumber, chilli, basil and thyme.

Bruichladdich Distillery, Isle of Islay
thebotanist.com
Price £35 / Quantity 70cl / ABV 46%

THE BOTANIST

FOUNDED IN 1881 on the Hebridean island of Islay by three brothers – William, Robert, and John Gourlay – Bruichladdich (pronounced 'Broo-k-lad-dee') distillery has suffered mixed fortunes over the years. It was considered state-of-the-art when it was built, but in 1994 it was eventually mothballed.

While wine merchant Mark Reynier's decision in December 2000 to buy the neglected production site was certainly a good one, the real stroke of genius was his decision to poach industry legend Jim McEwan from Bowmore.

It was Jim's idea to expand production to include white spirits and in 2012, after much consultation with local botanists, Islay-based Richard and Mavis Gulliver, the first batch of the Botanist Gin went on sale.

The product of many years of research and adaptation of the distillation techniques already employed to make whisky, the Botanist uses 22 botanicals that are locally sourced and hand-foraged.

The gin's name is not just a play on words, as the link between the team and their botanicals is key to the product's ongoing success.

Professional forager, James Donaldson, helps to source the plants from the hills, shores, and bogs of this Hebridean island.

These botanicals have not only been chosen for the flavour they impart but also for the fact that they are readily available and are in 'plentiful supply', thereby protecting Islay's delicate environment.

Jim McEwan then perfected this recipe on a Lomond still – charmingly dubbed Ugly Betty – which had been rescued from the old Inverleven Distillery in Dumbarton and transported to the island by barge. Since then, The Botanist has gone on to become one of the most successful of all Scottish gins, and in 2018 it even began outselling Bruichladdich's own whiskies.

.

BOTANICALS

Apple mint, chamomile, creeping thistle, downy birch, elder, gorse, hawthorn, heather, juniper, lady's bedstraw, lemon balm, meadowsweet, mugwort, red clover, spearmint, sweet cicely, bog myrtle, tansy, water mint, white clover, wild thyme, wood sage, orange peel, orris root, lemon peel, liquorice root, juniper berries, coriander, angelica, cinnamon and cassia bark

PREFERRED SERVE

Experimentation is best when it comes to enjoying this gin

PREFERRED GARNISH

Try adding some thyme and lemon balm, fresh rosemary picked from the garden or a sprig of mint

RECOMMENDED COCKTAIL
BOTANIST COLLINS

Ingredients
50ml The Botanist Gin
20ml lemon juice
25ml cloudy apple juice
10ml elderflower cordial
Soda
Apple fan, to garnish

Method
1. Add all the ingredients in a highball glass with ice.
2. Top with soda.
3. Garnish with an apple fan.

THE BOTANIST

The first thing you'll notice about the Botanist is its eye-catching bottle, which features the Latin names of all 31 of its botanicals. Yes, you read that right: 31.

Of these, 22 are locally foraged including apple mint, chamomile, creeping thistle, gorse, bog myrtle, heather and hawthorn. These are then added to nine traditional botanicals, including coriander, angelica and cassia bark, as well as the juniper to create, as expected, a rich and complex gin.

Described as having a 'mellow taste with a citrus freshness', The Botanist is one of the most versatile Scottish gins and, as the producers are keen to point out, is perfect for experimenting with. Have some fun with it!

LUSSA GIN

THE ISLE OF JURA is famous for many things – its iconic whisky, the Paps and, of course, for being the inspirational location in which George Orwell took refuge to write 1984.

The popular island, which lies north of Islay on the west coast of Scotland, now serves as inspiration to three female distillers who are using it as a base for one of the country's most remote gins.

In the summer of 2015, self-taught and originally working from a tiny 10-litre still in the kitchen of one of their houses, Alicia MacInnes, Claire Fletcher and Georgina Kitching took their adventurous spirit and love of foraging and used it to create a gin that they describe as a 'distillation of the island'.

The trio all live on Lussa Glen in Ardlussa on the north end of the island, distilling, bottling and labelling the gin themselves. A true adventurist's gin, Lussa captures not only the island itself but also reflects the environment in which it is nurtured.

The trio explore all possibilities in their hunt for the best botanicals, be that rowing out to sea to gather sea lettuce or climbing trees to harvest Scots pine needles. The botanicals they gather by hand are then frozen rather than dried.

And all of the 15 chosen for the recipe for their aromatic, zesty London Dry gin can either be grown or foraged on the island. The island's spirit of community means that many Diurach – residents of Jura to you and me – help to produce what's needed: in return for a bottle or two of the finished product, naturally.

Having recently expanded into selling in both Germany and the USA it seems the appetite for this exciting brand continues to grow, with the team adding a bramble liqueur and even Lussa Gin soap to their popular range.

LUSSA GIN

Using only botanicals that can grow on Jura or be foraged from its hills, bogs, coastline and woods, Lussa Gin features prominent Scottish ingredients such as

The Lussa Gin Distillery, Isle of Jura
lussagin.com
Price £40 / Quantity 70cl / ABV 42%

BOTANICALS
Lemon thyme, coriander seed, rose petals, lemon balm leaves, lime flowers, elderflowers, honeysuckle flowers, bog myrtle, orris root, juniper, water mint leaves, sea lettuce, Scots pine needles, ground elder and rose hip

PREFERRED SERVE
Walter Gregor's tonic water

PREFERRED GARNISH
A sprig of lemon thyme or mint with a frozen lemon wedge instead of an ice cube

RECOMMENDED COCKTAIL
LUSCIOUS LUSSA

Ingredients
50ml Lussa Gin
10ml lime juice
10ml elderflower cordial
3 leaves of lemon balm
1 sprig of lemon thyme
Soda

Method
1. Muddle all of the ingredients in a tall glass.
2. Fill with ice and top up with soda.

juniper – which grows wild on the west coast of the island – bog myrtle, Scots pine and rose hip.

As part of Lussa's drive to improve sustainability, the team have started a long-term project to propagate cuttings of Jura juniper and planted 500 seeds of certified Argyll juniper, as well as begun growing their own lemon thyme in polytunnels.

The result, distilled on a handcrafted Portuguese copper still – named Jim – is an exceptionally smooth, fresh gin filled with floral notes and slight hints of spice.

COLONSAY

BOTANICALS
Juniper, angelica root, calamus root, liquorice root, orange peel, orris root and coriander seeds.

PREFERRED SERVE
Serve over ice with a premium tonic and a garnish of either orange zest, or, most unusually, a slice of green chilli.

PREFERRED GARNISH
Orange zest or a slice of green chilli.

Isle of Colonsay, PA61 7YR
wildthymespirits.com
Price £45 / Quantity 70cl / ABV 47%

HUSBAND-AND-WIFE team, Finlay and Eileen Geekie, finished their self-build home on Colonsay in April 2016, and moved to the tiny Hebridean island not long after. The pair then founded Wild Thyme Spirits just a few months later. Colonsay Gin was launched in March 2017 and is described as a 'modern interpretation of the classic London Dry style gin' – meaning it's as delectable when sipped neat as it is as the central component of a cocktail.

Finlay explains that they chose to make this traditional style of juniper-led spirit so it would stand out from an increasingly crowded marketplace, which he finds to be 'brimming with flavoured gins infused with unusual botanicals'. Instead, Colonsay would be instantly, sharply recognisable as a gin.

As such, their recipe features botanicals considered typical for a classic London Dry gin, with the exception, of course, of the calamus root, which brings a fiery 'ginger-like rootiness' to Colonsay. In 2018 the Geekies moved production from Methven to the island and are confident that with the foraging rights they have on the Colonsay Estate, they will create newer expressions that fully showcase the island on which their new distillery is based.

Should any gin aficionado fancy sampling a slice of Colonsay life, then the Wild Thyme Spirits team also offer a 'Gin Lover's Retreat' which includes enjoying the 'remote Hebridean beauty' of the island, the choice of over two hundred gins from around the world, gin cocktails, a tasting and accommodation.

COLONSAY GIN

Described as a classic London Dry style gin, Colonsay is a juniper-forward gin, with strong pine notes upfront, followed by an earthy root sweetness. Coriander seeds lend warmth and depth to the gin, while orange peel brings subtle citrus and calamus offers a kick of fiery ginger.

The gin lives on long after the final sip and the lasting finish is that of the sweet roots. Inspired by Celtic folklore, the bottle's stunning label was created by South African illustrator Caroline Vos and shows the distillery's three 'Brownies' (a Brownie or Uruisg – pronounced 'oor-isk' – in Gaelic was a type of spirit that helped with household chores): ALVA, a delightful, red-haired maiden and twins Doughal and Ferghus, as well as the island that they call home.

KILORAN WAVES

Created by Andy Mil from Cocktail Trading Company in London.

Ingredients
50ml Colonsay Gin
1 tsp greengage jam
20ml seaweed and tea syrup*
5ml smoky Scotch (e.g. Ardbeg)
25ml lime juice
1 egg white
Salt, lime and an edible flower to
  finish

Method
1. Shake all the ingredients in a cocktail shaker
2. Fine-strain into a cocktail glass.
3. Garnish with a sprinkle of salt on top of the glass, spritz with a twist of lime and decorate with an edible flower.

To make the syrup
1. Add 1 tsp loose-leaf breakfast tea, 1 sheet (2.2g) of Nori seaweed and 500g caster sugar to 500ml of hot, not boiling, water.
2. Dissolve the sugar and leave to infuse for ten minutes before straining. This syrup will last in the fridge for two months.

THE ISLANDS

WILD ISLAND

89

BOTANICALS
Redcurrants, lemon balm, sea buckthorn

PREFERRED SERVE
A simple G&T, lots of good-quality ice, serving of Distiller's Cut and a good tonic water

PREFERRED GARNISH
A sprig of mint and a scattering of raspberries

Colonsay Beverages Ltd, Isle of Colonsay
wildislandgin.com
Price £39 / Quantity 70cl / ABV 43.5%

SITUATED JUST A short ferry journey from Oban on the west coast, the 'Sunshine Island' of Colonsay is tucked away between the better-known islands of Islay, Mull and neighbouring Jura.

It was on this tiny island, with a diameter of just ten miles and a population of about 130 people, that in 2007, after a night in the island's iconic hotel drinking very average beer, four friends David Johnston, Chris Nisbet, Keith Johnston and Bob Pocklington decided that they would give brewing on the island a go, launching Colonsay Beverages.

Arriving with a vast experience of the Scottish drinks industry, two new investors Keith Bonnington and Allan Erskine soon followed, leading to the decision to add a gin that would reflect the sights and smells of the flora as you walk around the island being added to their portfolio.

Launching Wild Island Gin in 2016, at the start of what would go on to be the current Scottish craft gin boom, they enlisted the help of Master Distiller Rob Dorsett of Langley Distillery in Hertfordshire to help cook up a recipe that would showcase the amazing botanical larder they had on their doorstep, whilst adding value to the local island economy.

In summer 2017, the first mini still arrived on the island and Chris began experimenting, under the watchful eye of Rob, with different fruits and flowers he could find around his croft.

This led to the first in their mini botanical series, which was such a success, they decided to add it to their core range of products in a 70cl bottle, with a 250-litre G-still now producing small-batch spirits for them on the island.

Chris now balances his gin making with being the island's chief firefighter, air traffic controller, house builder and crofter, but you can still visit the distillery for gin tastings, shopping and island chat with Sheena, his wife, six days a week.

Continuing with the Distiller's Cut range they have also released a honey gin, made with Colonsay wildflower honey, and an Islay cask-aged gin.

DISTILLER'S CUT GIN

A true showcase of the wonderful flora on Colonsay, in August the island is bursting with juicy redcurrants and it's these, combined with island foraged lemon balm, wild water mint,

---

RECOMMENDED COCKTAIL
BLACKBERRY MOJITO

Ingredients
25ml Wild Island Distiller's Cut
2 quarters of lime
Handful of fresh mint leaves
1 tsp brown sugar
6 blackberries
Mexican lime soda

Method
1. Muddle 2 lime quarters, a handful of fresh mint leaves, a teaspoon of brown sugar and 6 blackberries in a glass.
2. Add ice and 25ml of Wild Island Distiller's Cut.
3. Top with Mexican lime soda and finish with a sprig of mint.

---

meadowsweet, heather flowers, sea buckthorn and bog myrtle that come together to create Distiller's Cut Gin.

It's a gin they can be proud of, and rightly so with the extra challenges they face distilling on a remote island. Its refreshing mix of lemon and mint is followed by the sharp tang of ripe redcurrants and the lingering floral finish of the meadowsweet and heather flowers.

Tobermory Distillery, Isle of Mull
tobermorydistillery.com
Price £34 / Quantity 70cl / ABV 43.3%

90

TOBERMORY HEBRIDEAN

LYING OFF THE west coast of the mainland, is the Isle of Mull and its most prominent town of Tobermory, with its colourful houses serving as the inspiration for the island's first legal gin.

The only whisky distillery on the island, and one of the oldest in the country, is better known for its Tobermory single malt whisky and Ledaig heavily-peated malt but made its entrance into the white spirits market in summer 2019.

Thanks to the acquisition of an old John Dore & Co copper still from South Africa, which was built in the 1950s and was shipped to Scotland for refurbishment and installation, the distillery team were able to create their new gin.

Altering parts of the distilling site, including converting a storage area and old office, to accommodate the new equipment, allowed them to fit the full length of the still neck and create a lighter vibrant character in the final spirit.

Having established their first gin, the Tobermory team quickly followed up with their first limited edition, the Mountain Gin, which features Mull-grown and foraged ingredients such as tea, rowan berry, rose hip and wild heather.

Tobermory have also added an adult colouring book inspired by the dramatic scenery of the idyllic Hebridean island to their range, with ten carefully hand-drawn illustrations by talented artist, Lydia Bourhill.

BOTANICALS
Tobermory malt spirit, juniper, heather, Hebridean tea, heather, elderflower, orange peel

PREFERRED SERVE
Fill a balloon gin glass with copious amounts of ice and a good measure of Tobermory Hebridean Gin and add a premium tonic water

PREFERRED GARNISH
Sprig of thyme, fresh blood red orange slice, pinch of dried hibiscus flowers

COCONUT MOUNTAIN

Ingredients
50ml Tobermory Gin
50ml pineapple juice
25ml coconut water
10ml sugar syrup/gomme
2 dashes of Angostura Bitter

Method
1. Vigorously shake all ingredients in a cocktail shaker
2. Strain the mix into a coupe glass.

TOBERMORY HEBRIDEAN GIN

Hebridean by name, Hebridean by nature, Tobermory Gin is designed to reflect the island it calls home. Featuring the same 'fruity' character seen in the firm's whiskies, this colourful spirit is made using 13 specially selected botanicals, including a splash of new-make spirit from the Tobermory whisky stills.

Added to this is a pinch of Hebridean tea, grown locally on Mull, along with elderflower and sweet orange peel, to create a balanced gin with fruity and floral notes which roll around the tongue thanks to the malty mouthfeel of the added newly distilled single malt spirit.

Tobermory Harbour

91
WHITETAIL

THE FIRST NEW distillery on the Isle of Mull in over 220 years when it opened in 2019, the family-run Whitetail Spirits Limited were inspired by Europe's largest bird of prey when it came to naming their enticing spirit.

The sea eagle has become synonymous with Scotland's 'Eagle Island' after it was re-introduced to the western isles in the 1970s, with the powerful raptor – which was once a common sight across Scotland until they were hunted almost to extinction in the 1910s – quickly making its home on Mull.

One pair in particular, which made their home in the forest close to where the Mackay family live on the Tiroran Estate and are regular stars of BBC's 'Springwatch' programme, gave them the idea to call their first release in 2017, Whitetail Gin.

Springing from a desire to create a classic smooth London Dry style gin, the family-run brand, who first moved to the island from Edinburgh in 2004, first teamed up with Charles Maxwell of Thames Distillers in London to contract distil their first gin produced using botanicals sourced on the island.

However, it wasn't long before they could realise their dream of building a distillery on their small estate and hotel, Tiroran House, on the south-west of the beautiful island, which lies just an hour from Tobermory.

Their popular gin is now distilled by them using botanicals foraged from the island including sea kelp collected by hand from the shores of the nearby sea loch Scridain in spring, calluna heather and winter savoury from the land around the estate and even pine needles from the nearby forest where the famous eagles nest.

Following on from their original Whitetail Gin, the firm have just launched a brand-new range of gin liqueurs including late summer berry, vanilla and rhubarb, and aqua mint and lime.

Also proving popular are their juniper and seaweed scented candles which are created using the distillery's used

The Steadings, Isle of Mull
whitetailgin.com
Price £38 / Quantity 70cl / ABV 47%

Juniper, coriander, angelica, lemon, orange, sea kelp, calluna vulgaris (Scottish heather), winter savoury, pine needles

PREFERRED SERVE
Over plenty of ice with a classic tonic water

PREFERRED GARNISH
Wedge of grapefruit and a sprig of rosemary

RECOMMENDED COCKTAIL

WHITETAIL SUNSET

Ingredients
35ml Whitetail Gin
25ml Dry white wine
A squeeze of fresh grapefruit juice
15ml Maple syrup
Top with soda

Method
1. Put in a mixer with ice, stir until the outside of the mixer is cold.
2. Strain into a wine glass with ice.
3. Garnish with a grapefruit twist and a sprig of rosemary.

bottles by upcycling specialists Compass Candles and are available in regular runs from the distillery's online shop.

WHITETAIL GIN

Deliberately designed as a classic London Dry gin that tells the story of the place in which it is made, the team combine botanicals such as the aforementioned foraged ingredients in sea kelp, heather and pine with more traditional items like coriander, angelica, lemon and orange to create a spirit that's full of flavour and most importantly, juniper-forward.

Delivered in a visually appealing bottle with glass the colour of Mull's seas and sky (the chosen habitats of the island's sea eagles) and an etching by acclaimed designer Tom Lane of Ginger Monkey Design, Whitetail Gin comes in at a higher ABV than most Scottish gins at 47%.

Still surprisingly smooth – in the team's own words 'as gentle on the palate as a feather drifting down from the sky' – the vibrant gin is distilled on a handmade 200-litre copper still from Portugal and offers subtle notes of the coast, floral tones from the heather as well as that full-on juniper hit you'd expect from a classical style gin.

TYREE GIN

BOTANICALS
Made using 12 botanicals. 5 locally sourced ingredients: kelp, ladies bedstraw, angelica root, eyebright and water mint

PREFERRED SERVE
Over ice with Fever-Tree Mediterranean tonic

PREFERRED GARNISH
Slice of pink grapefruit

1A West Hynish, Isle of Tiree
tyreegin.com
Price £34.99 / Quantity 70cl / ABV 40%

PRISTINE WHITE BEACHES, aquamarine water, sunshine galore, you'd be forgiven for thinking Tiree, the most westerly of the Inner Hebrides, was in the Maldives rather than off the west coast of Scotland.

It's this stunning scenery, coupled with the fact the island – sometimes known as Tir an Eòrna (Land of Barley in Gaelic) – was once home to 'no less than fifty distillers', that inspired Ian Smith and Alain Campbell, who had both just returned to the island, to set up the first legal distillery there since 1801.

Desiring to preserve and promote the island's whisky heritage and revive distilling on the island, the Tiree Whisky Company Ltd were given approval to begin creating the Isle of Tiree Distillery in Ian's dad's old builder's workshop in 2019.

With strong links to both the island's community and the local music festival, the pair first released a 19-year-old Speyside malt whisky, dubbed the Cairnsmuir after a ship that ran aground off the shores of the island with its holds full of whisky, to raise awareness of their new spirits range.

Then in 2017, the pair launched their first gin (distilled at the time by Thames Distillers in London) made using botanicals from the island.

Now fully produced on the island that gives it its name (all be it with an older spelling), Tyree Gin is designed to truly reflect the landscape in which it is distilled. Kelp harvested from the clear blue waters provides sweetness as well as coastal salty flavours, while the floral, grassy and vanilla notes are provided by a range of botanicals from the machair ground inland from the shore.

The enterprising duo recently followed up their popular island gin with the Hebridean Pink Gin, which is made using juniper berries, raspberries and sweet peels.

RECOMMENDED COCKTAIL
TYREE RASPBERRY SOUR

Ingredients:
50ml Tyree Gin
Chambord
Sugar Syrup
Lemonade
Garnish with Blueberries

Method
1. Add all ingredients into a cocktail shaker with some ice and shake for a few minutes until the shaker becomes cold.
2. Strain into a coupe or martini glass, top up with the lemonade.
3. Garnish with the blueberries.

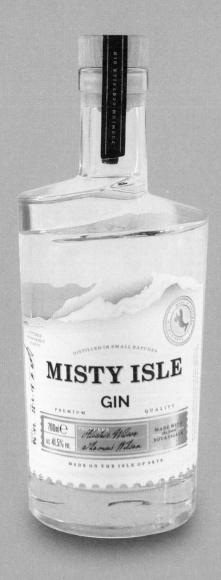

The Distillery, Portree, Isle of Skye
isleofskyedistillers.com
Price £36 / Quantity 70cl / ABV 41.5%

93
MISTY ISLE

CONSIDERED ONE OF the country's most beautiful regions, the west coast island of Skye is home to some truly spectacular and dramatic scenery. From its waterfalls and fairy pools to rock formations such as the Quiraing and the Storr, it enjoys phenomenal popularity with everyone from tourists to film directors.

Brothers Thomas and Alistair Wilson – Skye born and raised – combined their experience in the building trade and hospitality industry to open Skye's first gin distillery in Portree in 2016.

Naming themselves Isle of Skye Distillers, they built their distilling site in the back garden of a home that's been in the family for five generations.

The distillery is very much a family affair – the Wilsons work alongside other family members to produce their spirits. All of their gins are produced on their four copper stills, Chursty, Mairi, Euan and Finlay, while bottling and labelling is also completed on site.

The duo believe that provenance is everything, and sharing that abiding sense of belonging and community, along with the 'spirit of Skye', with the world became important to them. This led to their signature spirit, Misty Isle Gin, being launched in early 2017.

Deriving its name from the island's epithet, Misty Isle is designed to reflect Skye's stunning landscape, with the Old Man of Storr and the Cuillin mountain range offering particular inspiration.

This can be seen on the bottle itself: the label design and its multiple raised edges reflects the breathtaking mountains that form the island's backdrop, while the copper foiling gives a nod to their two original stills, Chursty and Mairi.

Growing from strength to strength, various award-winning expressions have since been added to their core range, with four new gins including Tommy's Gin, which features poppy seeds and honours the memory of their late father Tommy; their Pink Fruity Old Tom, Cill Targhlain, the newest London Dry, and the festive favourite Misty Isle Mulled

BOTANICALS
Juniper, orris root, liquorice root, black cubebs, coriander, grains of paradise, lemon peel, lemon verbena, cassia bark, angelica root and a secret Skye botanical

PREFERRED SERVE
Misty Isle should be served with a slice of orange peel and Scottish tonic, but mint lemonade makes a nice alternative to tonic. Another good serve is with a cinnamon stick and ginger ale

PREFERRED GARNISH
Orange peel, or a cinnamon stick

RECOMMENDED COCKTAIL
SHEPHERD'S DELIGHT

Ingredients
50ml Misty Isle Gin
25ml elderflower liqueur
25ml strawberry liqueur
Rose lemonade
Orange peel, to garnish

Method
1. Mix all the ingredients apart from the lemonade.
2. Shake over crushed ice and top with the rose lemonade.
3. Garnish with a twist of orange peel.

Christmas Gin, which uses all of the ingredients for a mulled wine, soaked in Amarone red wine before distillation.

They've also created a select range of Limited Edition Gins including the latest Halloween Spookily Spiced Gin.

Misty Isle Vodka, which was added in 2018, became the first vodka to be distilled on the island.

Still 100% independent and entirely funded by the Wilsons, the brothers have expanded to run a Gin School and Island Shop in the centre of Portree, and have designs on a further expansion with the distilling of whisky alongside their gins and vodka a key aim.

MISTY ISLE GIN

The gin itself combines the crystal-clear spring waters from the Storr Lochs with botanicals foraged on Skye. The balance of botanicals, including foraged juniper from various locations across Skye, means that Misty Isle is described as a juniper-forward gin.

It has an earthy mix of angelica root, liquorice root and orris root, with a good portion of citrus (lemon) and some spice from coriander and black cubebs as well as an intriguingly 'secret' ingredient, which can only be found at high altitude on Skye.

Old Man o Storr, Isle of Skye

Borodale House, Isle of Raasay
raasaydistillery.com
Price £34.95 / Quantity 70cl / ABV 46%

RAASAY

LYING JUST OFF the northeast coast of Skye is a tiny island that's fast becoming a spirits destination in its own right. The magical Hebridean Isle of Raasay, a short 25-minute ferry journey from Sconser, is filled with a diverse landscape of wild beaches, rocky coastal scenery and the flat-topped hill of Dùn Caan, which offers incredible views of the Cuillin mountains on Skye to the east, and Torridon on the mainland to the west.

Most interestingly for us, the island, which is rooted in centuries of illicit distilling, unveiled its first-ever legal spirits distillery in 2017, leading to the creation of not only its first whisky but also its first-ever gin too.

The first legal spirit from Raasay, this new handcrafted Scottish Gin is distilled in a Frilli copper pot still using an expert blend of ten botanicals.

Enlisting local botanist Dr Stephen Bungard and Fiona Williamson, MSc scholar at Heriot-Watt, who knows Raasay well having worked a summer

BOTANICALS
Juniper berries, rhubarb root, lemon peel, orange peel, coriander seeds, angelica root, liquorice root, orris root, cubeb pepper, and triple distilled Raasay spirit

PREFERRED SERVE
Poured over ice with a premium tonic ratio of 2:1 and garnished with orange zest strips

PREFERRED GARNISH
Orange zest strips

season at the distillery in 2018, the distillery staff worked to perfect the recipe that would go on to become their fledgling gin.

Founded by entrepreneurs Alasdair Day and Bill Dobbie, drinks firm R & B Distillers transformed an eighteenth-century Gothic villa hotel on the south-west coast of the island into what would

RAASAY '45

The Raasay Gin's take on the French 75 with the addition of rhubarb syrup to complement the rhubarb root in their gin.

Ingredients
40ml Isle of Raasay Gin
15ml fresh lemon juice
10ml rhubarb syrup
Champagne or sparkling wine, to
 top up
Lemon zest, to garnish

Method
1. Combine all ingredients except your chosen fizz in a cocktail shaker.
2. Add plenty of cubed ice to the shaker (covering the liquid) for a controlled dilution and to avoid a watery flat cocktail.
3. Shake for 15 to 20 seconds or until the shaker is frosty to the touch and fine strain into a chilled Champagne flute.
4. Top with your choice of fizz and garnish with lemon zest.

become not only the distillery but also a visitor centre and luxury guest accommodation for those visiting the island too. With a key focus on tourism, the transformed Borodale House may just be the only site in Scotland that provides luxury accommodation within the same building as a working distillery.

Their first-ever single malt was released at the tail end of 2020, with this inaugural release being snapped up by fans and selling out its entire 7,500 bottle run before it was even launched.

Distillery co-founder Alasdair Day states that Raasay's remarkable geology and their modern island distillery were key in inspiring both the creation and presentation of their exciting Scottish gin.

Teaming up with two other neighbouring distilleries on Skye (Torabhaig and Talisker) as well as the exciting Isle of Harris Distillery, they've launched the Hebridean Whisky Trail, introducing this newest whisky route that may in time come to rival Islay and Orkney for their island whisky destination crowns.

ISLE OF RAASAY GIN

Inspired by the unique geological strata of the island, the first thing you will notice about the Isle of Raasay bottle is its beautiful label featuring the various rock types that the natural spring water runs through.

The gin itself is made in one run for the full year, using a recipe with ten carefully curated botanicals including the rhubarb root, lemon and sweet orange peel, coriander seeds and cubeb pepper as well as a drop of their triple distilled Isle of Raasay malt spirit and that all-important spring water – sourced from their on-site well, Tobar Na Ba Bàine 'The Well of the Pale Cow'.

It is a complex but well-balanced dry gin that perfectly encapsulates the island on which it's made – this little taste of the Hebrides will have you dreaming of a sunny weekend on Raasay.

View to Cuillins on the Isle of Skye from Hallaig, Isle of Raasay

Isle of Harris Distillers, Isle of Harris
harrisdistillery.com
Price £40 / Quantity 70cl / ABV 45%

95

ISLE OF HARRIS GIN

THE LAST BASTION of land before you hit thousands of miles of the Atlantic on the way to Newfoundland in Canada is a breathtaking, rugged place like no other. Isolated and ethereal, it's perhaps not where you'd expect to find a gin distillery.

Described as Scotland's first social distillery, the Isle of Harris Distillers lies where North and South Harris meet and is one of only a handful of spirits production sites to be found on the Outer Hebrides.

It was built as part of a project to create production and tourism jobs for Harris and its community, and also to help stop population decline on an island that's arguably one of Europe's most beautiful.

With whisky the main target of the new spirits production site, few could have imagined how successful the gin, initially conceived as a keepsake for tourists to buy at the distillery, would become.

Created using hand-dived sugar kelp, on their small copper gin still known affectionately as The Dottach after 'a

BOTANICALS
Hand-harvested sugar kelp, Macedonian juniper berries, English coriander seed, cubebs (Javan pepper), bitter orange peel, angelica root, cassia bark, orris root and liquorice

PREFERRED SERVE
In a tall glass with several large cubes of ice, add a splash of a premium tonic (Walter Gregor's Scottish tonic is the in-house recommendation) and add a few drops of the distillery's own Sugar Kelp Aromatic Water if you really want to take this G&T to the next level

PREFERRED GARNISH
Red grapefruit

similarly fiery and feisty local woman', the gin comes in a bottle that's sure to be one of the prettiest you'll come across: wave-like patterns are etched upon its aqua blue glass in tribute to its coastal home and the seas of Luskentyre.

Each bottle is created with deliberate imperfections to make it easier to hold while pouring – this imaginative, meticulous team really have thought of everything.

So popular are these uniquely rippled bottles that the distillery had to introduce a rationing system in 2016 when the specialist Yorkshire factory that makes them struggled to keep up with demand.

As part of a recent drive towards being more sustainable and helping their fans to reuse their beautiful bottles, they've just launched a new subscription service that will see a recycled aluminium refill pack sent to their door each month.

ISLE OF HARRIS GIN

Isle of Harris Gin – launched in 2015 – was the distillery's first gin. To create the recipe for it, a consultant botanist selected sugar kelp, an ingredient popular in Japanese cookery, ahead of other island plants such as heather and bog myrtle.

The dry kelp brings a saltiness to the final spirit, but also, surprisingly, a sweet flavour. Managing director Simon Erlanger points out that the distillery only uses small quantities of the kelp, as its flavour can be very powerful.

Sweet and spicy, with a slight hint of that wonderful sea air, Isle of Harris is an enigmatic gin that is more than worthy of the attention heaped upon it.

Hushinish, Isle of Harris

CAPER CEILIDH

Ingredients
60ml Isle of Harris Gin
12.5ml Dolin Blanc Vermouth de Chambery
2 drops of pink grapefruit-infused saline solution
Caperberry brine
Red grapefruit to garnish
3 caperberries to garnish

Method
1. Make the grapefruit saline solution by soaking the zest of a red grapefruit in a simple saline solution (20mg salt to 80ml water).
2. Chill a coupe glass with ice.
3. Add 12.5ml of the vermouth to a mixing glass with ice.
4. Stir, then strain off and discard the vermouth.
5. Add 60ml of gin to the vermouth-rinsed mixing glass and ice. Add 5ml of caperberry brine, and 2 drops of grapefruit saline solution.
6. Stir the contents to chill thoroughly.
7. Fine-strain into the chilled coupe glass.
8. Serve with a garnish of 3 caperberries and a twist of red grapefruit.

North Uist Distillery, Western Isles
northuistdistillery.com
Price £38 / Quantity 70cl / ABV 46%

DOWNPOUR

AFTER SAMPLING LIFE on the mainland, Jonny Ingledew and Kate MacDonald felt the call to return to their native island of North Uist, where they returned with the dream of creating outstanding artisan spirits crafted to accurately 'capture the spirit of Hebridean Island life'.

Inspired by their stunning surroundings, they set up the North Uist Distillery, with Jonny, who has a Masters in Brewing and Distilling, taking on the role of head distiller and Kate, creative director, on the north-west of the island.

It was from here that they released their first gin, Downpour, in April 2019; the first legal spirit produced on Uist is a 'bold-flavoured' premium gin created using wild Hebridean heather which is foraged from across the island and, most interestingly, embraces the spirit's natural clouding effect.

Quickly growing popular with gin fans, it wasn't long before they outgrew their original set-up and decided to move to a more spacious location. Incredibly,

BOTANICALS
Juniper, coriander, angelica, orris, lemon, orange, grapefruit, clove, cassia, cubebs, cardamom, wild heather flowers

PREFERRED SERVE
Over ice with Fever-Tree Indian tonic water

PREFERRED GARNISH
Lemon and rosemary or lemon and wild thyme (when in season)

not only did they find one in what was a former farmstead on Benbecula, but the new site also just so happened to be linked to Bonnie Prince Charlie and his escape to Skye in the aftermath of Culloden.

With a desire to be at the heart of the community on the islands, Kate explained that due to the fact they both grew up on Uist, it's important to them that each of

THE ISLANDS

DOWNPOUR GINQUIRI
(Wheesht Bar)

Ingredients
35ml Downpour Scottish Dry Gin
25ml grapefruit
20ml lemon
1 tbsp honey
1 sprig of rosemary

Method
1. Shake all the ingredients with ice in a cocktail shaker including the fresh rosemary.
2. Strain into a chilled glass.
3. Top with tonic (optional).

their gins is distilled, bottled and labelled on the islands, to enable the business to have a long-term benefit for the people there, with Benbecula and Uist at the heart of all their decision-making as they develop the brand.

The pair also recently announced plans to make whisky at their new distillery. With a desire to put the islands on the 'Whisky Map', they say they plan to make whisky using a grain that local crofters produce, Bere barley, which is one of the oldest cereals grown in the UK and has a great flavour for making

Scotland's national spirit.

Keen to work with local crofters, they hope to create a supply chain of Bere barley on the islands, with the intention of one day being able to malt the barley, allowing for full grain-to-glass production in the new premises, Nunton Steadings.

As well as the Downpour Scottish Dry Gin, they have also created a Sloe & Bramble Cask Aged Gin, a Pink Grapefruit Gin and even a pre-made cocktail in the Oak Aged Negroni.

DOWNPOUR GIN

Designed to fully embrace the spirit's natural clouding effect, the result of the 'downpour' of essential oils from the botanicals coming out of the solution from the alcohol concentration and the temperature dropping in the glass, this intriguing gin is clear in the bottle but will go cloudy when it is mixed with tonic and ice.

Known in the industry as 'louching', Jonny and Kate say this 'flavour cloud' is an indication of the strong, bold flavours inherent in this classic juniper Scottish dry gin.

Featuring botanicals like lemon, orange, grapefruit, clove, cassia and cubeb berries, Downpour is made using wild heather flowers sourced from around the islands. The pair offer a barter system to residents and visitors who harvest the heather flowers for them when they're in season, providing gin in return.

With big hits of citrus from the lemon and grapefruit, the heather flowers add sweet, honeyed notes that combine with the peppery kick from cubeb berries and the forest freshness of the juniper to create a deep and complex gin.

Croft community, North Uist

Isle of Barra Distillers Ltd, Isle of Barra
isleofbarradistillers.com
Price £37 / Quantity 70cl / ABV 46%

BARRA ATLANTIC

LYING OFF SCOTLAND'S west coast and offering the last stop on the Atlantic before you hit North America, the tiny island of Barra will soon be home to its first-ever dedicated whisky and gin distillery.

Filled with a rich history, Barra, along with neighbouring Isle of Vatersay, is the most westerly inhabited island in the UK and is the home of Barra Atlantic Gin.

The remote location where Sir Compton MacKenzie was inspired to write the novel *Whisky Galore*, the stunning natural beauty of this island, which is nicknamed Barra-dise, also provided the inspiration for husband-and-wife team Michael and Katie Morrison to create a spirit that reflected their beautiful home.

First launched in 2017, the pair initially tapped into the expertise of a distillery in London to perfect their juniper and carrageen seaweed recipe, due to the prohibitive start-up costs of establishing a full spec distillery on the remote island from the get-go.

However, it wasn't long before Barra's

BOTANICALS
Carrageen seaweed, liquorice root, lemon peel, juniper, berries, angelica root, orris root powder, mint, Quebec peppers, lemon balm, elderflower, dried peppermint, coriander seeds, orange peel, cassia bark, camomile flowers

PREFERRED SERVE
Add ice, gently squeeze a pink or red grapefruit across the ice, just adding a few drops. Pour a 50ml measure of Barra Atlantic Gin then add a splash of Walter Gregor's premium Scottish tonic water and garnish with your wedge of grapefruit. Give a generous stir with a spoon

PREFERRED GARNISH
A wedge of grapefruit

first legal distillery, which they opened in May 2019, was joining the island's list of attractions that include a tiny airport with the world's only commercial beach runway, a 'Castle in the Sea' and snails

THE ISLANDS

that you'll find on restaurant menus in the capital.

Perfecting their Barra Atlantic Gin on their 300-litre copper gin still, which was custom built by Forsyths in Rothes and has since been named Ada, they've since unveiled plans to add a whisky to their spirits portfolio.

Providing a new home for 'Ada' as well as allowing them to create a single malt whisky and offer a visitor centre for guests to the tiny Hebridean island, this new £5m purpose-built distillery is expected to also create at least 30 new jobs for the local population of just 1,174 inhabitants.

Michael and Katie plan for the plant to be powered by renewable energy, and, on top of a build that will be constructed with sustainable materials, they hope to develop a green travel plan that will limit the number of visitors driving to the site.

Best of all, Katie confirmed that a dedicated café and bar will focus on local produce; from freshly caught local seafood and shellfish to local lamb and beef, adding that it will be a place that will 'celebrate the riches on Barra's doorstep'.

Isle of Barra

BARRA ATLANTIC GIN

Barra's first gin joined a growing list of food and drink exports alongside its now-famous snails, which have previously taken pride of place on the menu at top chef Fred Berkmiller's popular French cuisine restaurants in Edinburgh.

With the arrival of their custom-made still from world-famous Scottish coppersmiths Forsyths in November 2020, Michael and Katie were able to scale up production for the recipe they'd brought in-house, following the purchase of their original smaller still in May 2019 until their new copper still was installed in November 2020.

Describing the sight of seeing the first spirit trickle out of the still as 'emotional', the duo has placed a focus on key botanical Carrageen seaweed, a type of algae sourced naturally – and hand-picked – from the shallow waters surrounding the island.

Designed to 'perfectly encapsulate a taste of the island', the resulting gin is made using 17 botanicals and has a minerality with briny seaside notes that combine with the citrus and juniper flavours to create a deliciously balanced gin that's filled with floral, sweet, herbal and salty notes that will remind you of an afternoon spent on a windswept golden beach.

THE ISLANDS

The Orkney Distillery and Visitor Centre, Kirkwall, Orkney
orkneydistilling.com
Price £36.99 / Quantity 70cl / ABV 43%

KIRKJUVAGR
{KIRK-U-VAAR}

HAVING PREVIOUSLY BEEN home to two world-famous whisky distilleries and a hugely popular brewery, Orkney has embraced the world of modern Scottish spirits in a big way.

The past few years have seen the introduction of several award-winning gins and even a rum. One of the most prominent of these is Orkney Distilling Limited, which was founded in January 2016 by husband-and-wife team Stephen and Aly Kemp, and whose first product is the tongue-twisting Kirkjuvagr Gin – pronounced 'kirk-u-vaar'.

Created following a long period of working with consultants on research and experimentation, Kirkjuvagr's recipe was inspired by a local legend that spoke of a variety of angelica brought to the islands by Scandinavian seafarers centuries ago.

Still found growing wild on the Orkney island of Westray, this angelica was chosen as one of the defining botanicals in a recipe featuring other local botanicals, including ramanas rose, burnet rose and borage, as well as that most Orcadian of ingredients, traditional bere barley.

Each of these Orcadian botanicals are specially grown or hand-picked for the Orkney Distillery by the Agronomy Institute of the University of the Highlands and Islands, at their site overlooking Kirkwall Bay.

They have even developed a strain of calamondin oranges to add a kick of citrus to the signature gin. Originally contract distilled at Strathearn Distillery in Perthshire, the production of Kirkjuvagr and its sister gin Arkh-Angell (Storm Strength Edition – 57%) moved home in 2018 to the island's main town.

There you'll find the company's bespoke distillery, complete with two 200-litre copper stills, and a visitor centre on the harbour front, featuring a coffee shop as well as a gin bar.

This remarkable facility is the perfect excuse for any gin fan to visit the island; it offers a gin-making experience where fans of the Orcadian spirit can spend all

BOTANICALS

Juniper, angelica root, aronia berries, borage, cassia bark, coriander seed, holy thistle, lemon peel, liquorice root, marshmallow root, milk thistle, nutmeg powder, orange peel, orris root powder, rose hips (ramanas and burnet rose), calamus root, heather flowers and finally some bere barley

PREFERRED SERVE

Over ice with your favourite tonic

PREFERRED GARNISH

Orange peel

RECOMMENDED COCKTAIL

PAMPLEMOUSSE

Ingredients
35ml Kirkjuvagr
25ml Pamplemousse (grapefruit liqueur)
50ml pink grapefruit juice
Dash of grapefruit bitters
Sugar syrup to taste
Dehydrated pink grapefruit, to garnish

Method
1. Mix the ingredients together and pour into a coupe glass.
2. Garnish with the pink grapefruit.

day at this unique distillery, learning the secrets of crafting their own gin on a miniature still and receiving a bottle to take home with them.

The Orkney Distillery is also home to four other gins; the lighter, sweeter Harpa, named for the arrival of spring; the winter edition Aurora, named after the renowned Aurora Borealis; and Beyla, their Old Tom style gin which is named after the Norse Goddess of bees and fertility.

Finally, the most recent gin to be added to their series is a collaboration between Orkney Distilling Ltd and the Northwest Expedition Team.

Developed to help fund a team of adventurers and ocean rowers aiming to be the first to row the Arctic's Northwest Passage, known as the Last Great First, it will also support marine conservation through a partnership with Big Blue Ocean Cleanup.

KIRKJUVAGR GIN

Keen to retain a connection to Orkney's Viking heritage, the Orkney Distillery chose 'Kirkjuvagr' for the name of their gin. It's the Norse word for 'Church Bay' – the 1,000-year-old Viking name for their hometown of Kirkwall – and reflects the distillery's stellar setting and the home of those ancient seafarers who claimed the islands in their own name in those pre-medieval days.

Kirkwall Harbour, Orkney

A modern gin that's 'quite simply, unmistakably Orcadian', Kirkjuvagr is a complex spirit with notes of sweetness, salt and spice all following the initial floral hit. The packaging features the rolling waves of the seas around Orkney and the Vegvisir, or Viking compass, reinforcing that strong connection with the Norse seafaring tradition. Should you be looking for a novel way to dispense Kirkjuvagr, the team sell their own hand-crafted gin dispensing taps. What could be better?

THE ISLANDS

SEA GLASS

BOTANICALS
Juniper, cucumber, mint, lavender, lemon
verbena leaf and tarragon

PREFERRED SERVE
Ice and Franklin & Sons tonic water

PREFERRED GARNISH
Blueberries or a slice of kiwi

Deerness Distillery Ltd, Orkney
deernessdistillery.com
Price £36 / Quantity 70cl / ABV 43%

THE RUGGED BEAUTY of Scotland's islands is complemented by their wonderful natural larders and the entrepreneurial spirit of a people who have no choice but to look beyond the ready availability of items found in the big cities of the mainland.

Head north to Orkney and the north-eastern coast of the main island and you'll find one of the country's most thrilling new producers. Launched in 2017, Deerness Distillery aimed to expand the island's distilling scene from the production of a globally renowned single malt whisky at Highland Park and Scapa, to an exciting range of craft spirits such as gin, vodka and, incredibly, small-batch rum and whisky!

Founded by Stuart and Adelle Brown, Deerness is the first new distillery on Orkney for 130 years. The family-run business produced its first spirits – Sea Glass Gin, Scuttled Gin, Vara Pink Gin, Into the Wild Vodka and Orcadian Moon coffee liqueur – on their three Portuguese copper alembic stills, which are named Walt, Zing and Matilda.

Construction of this innovative new site began in 2016 with six gins originally produced before they launched a year later. Each of the six was put to the taste test before Sea Glass was chosen as the flagship juniper spirit.

Their site boasts a beautiful shop and tasting area full of their products, merchandise and local arts and crafts, as well as a wee takeaway café for tea, coffee and snacks.

Their innovative polytunnel system allows them to grow a variety of botanicals on-site in a bid to protect the environment of this wild, spectacular island.

SEA GLASS GIN

Named for the glass found while combing the island's exhilarating beaches, Sea Glass Gin is unique for a few reasons.

The first new gin to be solely distilled on the island of Orkney, it is made using tarragon instead of coriander, which founder Stuart believes brings a smoother flavour and gives their gin a truly distinctive profile.

Cut with pure Orcadian water, Sea Glass is made on Matilda – the family's gin still – and is described as tasting of spices, citrus and juniper.

RECOMMENDED COCKTAIL

SEA PINK

Ingredients
45ml Sea Glass Gin
15ml Into the Wild Vodka
Crushed ice
45ml blueberry juice
45ml Franklin & Sons tonic water
Blueberries and lemon rind to
  garnish

Method
1. Coat the rim of a 150ml martini glass with lemon juice and sugar.
2. Mix all the ingredients together and garnish with blueberries and a lemon rind spiral.

Saxa Vord Distillery, Shetland Islands
shetlandreel.com
Price £34 / Quantity 70cl / ABV 43%

SHETLAND REEL

SCOTLAND'S MOST NORTHERLY distillery can be found on its most northern collection of islands.

Rugged, remote and beautiful, the island of Unst, Shetland, is the location of the former RAF Saxa Vord base, which is now home to the Shetland Distillery Company, and more recently the Shetland Space Centre, which will be the UK's Pathfinder Vertical Launch site for small satellites.

The distillery was launched in 2014 by two couples, Debbie and Frank Strang and Wilma and Stuart Nickerson, and blends both couples' desires to establish a company producing top quality spirits embedded in the local community.

The Strangs played a major role in the revival of Saxa Vord after the RAF left, and Nickerson is an industry veteran who had previously revived the Glenglassaugh Distillery in Aberdeenshire before selling it to the BenRiach group.

The quartet now distils all of their gins on the island, the unique location of which means that this most northerly of distilleries is also the most easterly, as well as being the most remote.

In fact, it is closer to Bergen in Norway than to Aberdeen, and closer to the Arctic Circle than it is to London.

Launched initially with both gin and whisky in mind, the distillery debuted with Shetland Reel, which was inspired by the island on which it is made.

It then followed up with Ocean Sent, made in honour of the seas around Unst, and Simmer Gin, created in celebration of Shetland's famously drawn-out summer twilight, which has the delightful local name of 'Simmer Dim'.

In celebration of the island's culture and rich ties to the past, the team also created a selection of small-batch, cask-aged, Navy Strength Up Helly Aa gins to celebrate the infamous Shetland fire festival and pay homage to the island's connection to the Vikings.

Whisky also plays its part with a malt whisky blended for them on the mainland, then shipped to the distillery in casks to be matured on Shetland for up to six months. It is reduced to bottle strength with local water and bottled.

Each recipe for the gins goes through a three-stage process.

It's first developed in a small glass still, before being perfected in a small copper still, then finally produced on the full-size still. Where possible, each gin has a signature ingredient that ties it to Unst, or a focus on one of the many special Shetland connections.

In the case of the Ocean Sent, its bladderwrack seaweed is harvested from the shoreline, while the original

BOTANICALS
Juniper, coriander, angelica root, orris root, cassia bark, almond powder, orange peel, lemon peel and apple mint

PREFERRED SERVE
Mix with Fever-Tree Original tonic and serve in a rocks glass with plenty of ice

PREFERRED GARNISH
A wedge of pink grapefruit, sprigs of mint and lavender

Saxa Vord, Unst, Shetland Islands

Shetland Reel uses locally harvested apple mint.

More recently Filska, the Shetland dialect word for flighty and high spirited, lends its name to their new pink grapefruit gin, while Wild Fire was produced to coincide with Ann Cleeves's final book upon which the Shetland drama series is based and contains locally grown sorrell to add a little spice and heat.

SHETLAND REEL ORIGINAL

Created to provide a unique snapshot of the island on which it is made, Shetland Reel original gin features nine key botanicals.

The most important of which – other than the juniper – is the apple mint, grown and harvested locally and dried before being added to more traditional ingredients like coriander seeds, angelica root and citrus peel.

This creates a gin that's traditional in style but has a citrus sweet flavour and refreshing yet subtle notes of mint.

RECOMMENDED COCKTAIL
SHETLAND SOUTHSIDE

Ingredients
50ml Shetland Reel
25ml pink grapefruit juice
10ml simple syrup
10ml lemon juice
6 to 8 mint leaves
Soda
Grapefruit rind, to garnish
1 sprig of mint, to garnish

Method
1. Muddle the mint and gin together in a shaker, add the remaining ingredients, top with ice and shake until cold.
2. Double-strain into a tall glass filled with ice and top up with soda.
3. Garnish with grapefruit rind and a mint sprig.

COCKTAIL INDEX

THANK YOU TO . . .

WRITING A BOOK isn't easy and it would have been even harder without the help and support of the following people (in addition to the team at Black & White Publishing; thanks for your infinite patience, skill and amiable nature).

With thanks to: Julia Fletcher Smith, for being my anchor and keeping me grounded when I felt overwhelmed, for always offering a word of encouragement, and being on hand to share the occasional gin with me.

My mum for providing me with a love of reading and writing; my dad for giving me the confidence to speak to people about their passions; and my stepdad for always being there with a beer.

David and Marjorie Smith, for advice and for plenty of opportunities to enjoy a gin or two.

Natalie and Martin Reid at the Gin Cooperative, for always being on hand to help with contacts, advice, information and especially a much-needed photograph – but also for their top support for the industry itself.

Dave Broom, for being a gin guru and taking the time to chat even when on holiday on Islay!

James Sutherland for being my brilliant co-host on our TalkGin podcast and giving me an ear to bend whenever I needed it.

Jamie Shields at the Summerhall Drinks Lab, for listening to my crazy requests, offering advice and helping me get my head around the stranger cocktail terminologies.

Jonathan Engels, for unravelling the world of Scottish juniper for me and for teaching me why so few Scottish producers can use it sustainably.

Blair Bowman, for being out there posing the question of what actually constitutes a Scottish gin, and driving the debate forward.

Geraldine Coates, for politely putting up with my pestering.

Kirsty Black, for chat about distilling, flavours and botanicals.

And, finally, to all the producers included in this book. Without you, this fledgling industry would be nothing, so thank you for making it what it is and for producing your amazing gins.

IMAGE CREDITS

PHOTOGRAPHS OF GIN BOTTLES BY
BEN BROWN (BENBROWNPHOTOG@GMAIL.COM),
EXCEPT FOR IMAGES ON PAGES LISTED BELOW.

viii, 2, 5, 10, 11, 12, 14-15, 67, 93, 105, 149, 195, 240, 256, 265 shutterstock

18, 20, 40, 48, 56, 60, 68, 96, 126, 130, 134, 143, 170, 178, 186, 190, 192, 196, 208, 212, 228, 232, 234, 246, 254, 270, 272 Black & White Publishing

16 Courtesy of Crafty Distillery

27 Courtesy of Lucky Lion Ltd

32 Courtesy of McLean's Gin

44 robertharding/Alamy Stock Photo

58 Bailey-Cooper Photography/Alamy Stock Photo

66 Rod Sibbald/Alamy Stock Photo

92 Courtesy of Edinburgh Distillery

100 Courtesy of Secret Garden Distillery

106 David Noton Photography/Alamy Stock Photo

108 Courtesy of McQueen Gin

137 Courtesy of Verdant Distillery

150 Maritxu22/Alamy Stock Photo

153 Scott Sim/Alamy Stock Photo

201 ESPY Photography/Alamy Stock Photo

202 Courtesy of Thompson Gros Distillery

205 Courtesy of North Point Distillery

216 Courtesy of Sutor's Gin

222 John Alexander Howie/Alamy Stock Photo

249 mauritius images GmbH/Alamy Stock Photo

253 Scotland by Jan Smith Photography/Alamy Stock Photo

261 Lynne Evans/Alamy Stock Photo

264 ARV/Alamy Stock Photo

266 Courtesy of The Orkney Distillery

269 John Peter Photography/Alamy Stock Photo

274 geogphotos/Alamy Stock Photo.jpg